CW00796318

'Will you not covet such power as this...?'

JOHN RUSKIN

CEB 994C

Stuart Gibbard

Published by
Old Pond Publishing

First published 2001

ISBN 1 903366 17 8

A catalogue record for this book is available from the British Library

Published by

Old Pond Publishing
104 Valley Road, Ipswich IP1 4PA,
United Kingdom

www.oldpond.com

Cover illustration:
Doe Triple D tractors on demonstration in 1961.

Frontispiece illustration:
A Doe 130 at work, ploughing in mustard seed.

Cover design and book layout by Liz Whatling
Printed and bound in Great Britain by
Butler & Tanner Ltd, Frome and London

Contents

Acknowledgements

This is the history of a family company and I am very grateful to Ernest Doe & Sons Ltd for allowing me to tell that story and making this book possible. In particular, I would like to thank Alan Doe, the current chairman, for his time and knowledge as well as access to the company archives. He has read my manuscript and was kind enough to offer very encouraging and positive comments as well as suggesting corrections and improvements where needed.

I could not have written this book without the assistance of Alan's secretary, Eileen Hockley, a very organised lady who had all the information, photographs and records at her fingertips. I doubt whether there is anyone else in the company that knows more about Triple D tractors than she does. She has spent a great deal of time sorting out what I needed for my research and her help was invaluable.

Eileen's husband, Derrick Hockley, although now retired, worked with Ernest Doe & Sons as apprentice, service engineer and salesman, and demonstrated the Triple D tractors sold onto farms in the U.K. He is recognised as the leading authority on the company and its products, and in particular, the Triple D and 130 tractors.

Even though Derrick has not been in the best of health recently, he still made me welcome and recounted his experiences with both patience and enthusiasm. I am very grateful to him for allowing me to share his knowledge and for kindly offering the use of his personal photograph collection. We spent many hours together or talking on the telephone and I am sure that we left no stone unturned or nut and bolt undone!

It is not the first time that Alan Doe, Eileen and Derrick have helped with one of my books, nor do I hope it will be the last. Alan and Eileen have attended several of my book launches, while I regularly meet up with Derrick at vintage shows across the country. This is the ideal opportunity to thank them for their continued support.

I must pay a special tribute to George Pryor, the man who was the inspiration behind the Triple D. George was kind enough to spend time with me going over the early development work and his reminiscences form an important part of this book for which I offer my sincere thanks.

I learnt a lot from Alwyn Blatherwick who runs the Doe Owners Club. Alwyn is a Triple D enthusiast and an expert on Doe tractors. He is a mine of information and I am very grateful for his time and assistance. I would also like to thank Kelvin Martin, the country's leading collector of Doe memorabilia, for his enthusiastic help, and Ronnie Crow, Doe's photographer during the Triple D period, for the loan of his negatives.

I am also indebted to Rob Atkinson, James Baldwin, John Blackbeard *(Arable Farming)*, Stephen Burtt, Bob Cavill, Trevor Clark, Philip Cook (Lord Rayleigh's Farms), David Cousins *(Farmers Weekly)*, David Fisher, Brian Foot (Frank Foot & Son), Rita Franklin (Lord Rayleigh's Farms), Nigel Ford, Steve Haylock, Peter Love, Peter Moore, Charlie Norman (Agrimac), Anna Oakford (Spindrift Photographic), Andrew Streeter, Peter Tewson (Agrimac), Michael Ward (Agrimac), Johnny Weal, Jan Wiggers and Paul Wylie. They have all been kind enough to provide information or photographs for either this book or my previous research on Doe tractors.

The usual thanks go to Roger Smith (publisher), Julanne Arnold (editor), Liz Whatling (designer), Lesley Smith (indexer) and Sue Gibbard (wife). They have all done such an excellent job that I have to be careful not to take them for granted.

Preface

Ernest Doe & Sons built just over four hundred and sixty tractors in less than ten years. Normally, such a short production run would not warrant a book of its own. But the Doe tractor was special, and so was the company that built it.

The Doe Triple D became a legend in its own lifetime and was one of the most unorthodox machines ever to appear on the market; it had two engines, four-wheel drive and could articulate through nearly 90 degrees. For the time, its performance was phenomenal; it was a sensation when it appeared in 1958 and it remains a sensation today with a following like no other tractor.

I first touched on the Triple D in one of my early books, *Ford Tractor Conversions*. That volume is long out of print, but it still creates interest and I regularly receive calls from enthusiasts asking about it and the chapter on Doe in particular. I felt that it was about time that I renewed my acquaintance with both the company and its famous tandem tractors.

I once owned a Triple D for several years so I know all its foibles, but thc machine still fascinates me. I wanted to expand on my original research, and was determined to discover as much about Doe as possible. The result, I hope, will go some way towards satisfying the Triple D enthusiasts, while creating a new following among those yet to be entranced by this enigmatic machine.

It was impossible to tell the complete story of Doe tractors without delving deeper into the history of the company. It too was unique; Ernest Doe & Sons evolved from a small village blacksmith's into one of the largest and most successful retail machinery organisations in the UK without forsaking its family traditions or values.

Space has precluded any more than a brief insight into the background and workings of the company. It deserves far more, but my main concern was with the machines that it built. For a fuller history of both Ernest Doe & Sons and the Doe family, I can heartily recommend Alan Doe's own book, *A Century of Service*, published in 1998 and still available through the company.

The Triple D, the 130 and the other Doe tractors account for only about ten years of the company's over 100-year history, but their importance must not be denied. They are the products that gave Ernest Doe & Sons worldwide recognition. The Doe name has a following as far afield as Holland, Germany, the USA and Australia. Ask tractor enthusiasts from these countries to name any other East Anglian machinery dealership ...!

Stuart Gibbard

July 2001

The author aboard his Triple D, taking part in the local village parade in 1990.

Foreword

By Alan E Doe, Chairman, Ernest Doe & Sons Ltd

I was very pleased to be invited by Stuart Gibbard to write the Foreword for his latest book, *The Doe Tractor Story*.

The company's success is very much due to the hard work, dedication and loyalty of the men and women of the firm during the past hundred years. Ernest Doe & Sons was built on principles and practices that are still valid today and I am proud that the company has survived favourably for four generations.

The story of the Triple D and Doe 130 tractors is the chief content of the book and credit for them must go to George Pryor who had the original idea and to my father, Ernest Charles Doe, who had the foresight to put them into production. Last but not least, credit must go to the often forgotten Charles Bennett, who engineered the tractor and was works manager until his premature death in 1963 – without him it might never have happened.

Plans were made and passed for a new factory of 18,000 sq ft at Ulting, but in the end it was not needed and therefore never built.

It never ceases to amaze me how much interest there is in our tractors after all these years and the prices they command. There is even a Doe Owners Club run by Nottinghamshire collector Alwyn Blatherwick.

I congratulate Stuart on the many painstaking hours of research he must have undertaken to produce this book. He has come up with some facts that I had either forgotten or not known!

ALAN E DOE

June 2001

Alan E. Doe, the chairman of Ernest Doe & Sons Ltd., with one of his latest acquisitions - a Triple D tractor that cost him a small fortune compared to the original price!

ERNEST DOE & SONS LTD
ERNEST DOE & SONS LTD

Chapter 1

Ernest Doe & Sons

Fordson tractors exhibited by Ernest Doe & Sons at the 1949 annual ploughing match held at Boreham Airfield. A County Full-Track crawler is second from the left, while the E27N Major on the extreme right is fitted with a JCB Majorloader.

Ernest Doe & Sons Ltd is a large agricultural and construction machinery dealership based at Ulting, near Maldon in Essex. The company is a highly successful retail organisation operating across most of East Anglia and south-east England. It is a proud family firm, steeped in tradition with its roots dating back to the nineteenth century.

The company's founder, Ernest Doe, was born in 1876, the second son of Charles Joseph Doe, a miller from Terling in Essex. Once Ernest was old enough to work, he took his turn in the mill until his father felt that it was time for the boy to learn a trade of his own.

In March 1893, after Ernest had turned seventeen, Charles Doe arranged an apprenticeship for his son with the local blacksmith, Charles Wood, of Grays Farm in the nearby village of Hatfield Peverel. Wood's blacksmith's shop was at Ulting, a few miles south-east of Hatfield Peverel on the road to Maldon. The term of Ernest's apprenticeship was four years, and his starting wage was six shillings per week.

The young apprentice evidently made an impression on his master, because when Wood wanted to retire five years later, he offered to lease the blacksmith's shop to Ernest who took over the business with effect from 24 June 1898. Doe's present depot at Ulting stands on the site of the original blacksmith's shop, affectionately known locally as Doe's Corner.

It was often said that Ernest Doe established a high reputation through 'good, old-fashioned hard work and straight dealing', leading to close connections with the local farmers, including James Havis of Southlands Farm, whose daughter he married in 1901. Ernest and his wife, Alice, had three sons, Ernest Charles, Hugh and Herbert Walter. As the business prospered, Ernest Doe bought the freehold of the blacksmith's shop and acquired the neighbouring farm. He built a house, Hill View, next to the workshop in 1907.

All three of Ernest's sons were expected to do their bit in the blacksmith's shop, even while still at school. The eldest, Ernest Charles, was keenly interested in tractors and felt that they were the way forward for the future. He persuaded his father to attend the Ministry of

Ernest Doe (third from the left) is seen shoeing a horse outside his blacksmith's shop at Ulting in 1899. He had taken over the business the previous year and quickly established a good reputation for hard work and honest dealing.

The blacksmith's shop at Ulting in 1908. Ernest Doe stands on the extreme right of the group. He both repaired machinery and acted as agent for a number of different makes of implements.

Munitions disposal sales where they purchased second-hand Fordsons that had been used during the First World War and sold them on to local farmers.

The major tractor trials were another source of inspiration; Ernest took his eldest son to the Lincoln trials in 1919 and again the following year. Ernest Charles wanted to buy the most powerful tractors at the trials, but his father curbed his enthusiasm and insisted that they would only consider the best. They were impressed with the performance of the cross-motor Case tractors, which were awarded a gold medal at the 1920 Aisthorpe trials, and wasted no time in securing an agency for these American machines.

Ernest Charles married in 1927, and his son, Alan Ernest, was born two years later. The family business was flourishing and took on the agencies for Fordson

The founder, Ernest Doe, who traded as the sole proprietor until Ernest Doe & Sons was formed in 1937. He remained senior partner of the firm until 1947.

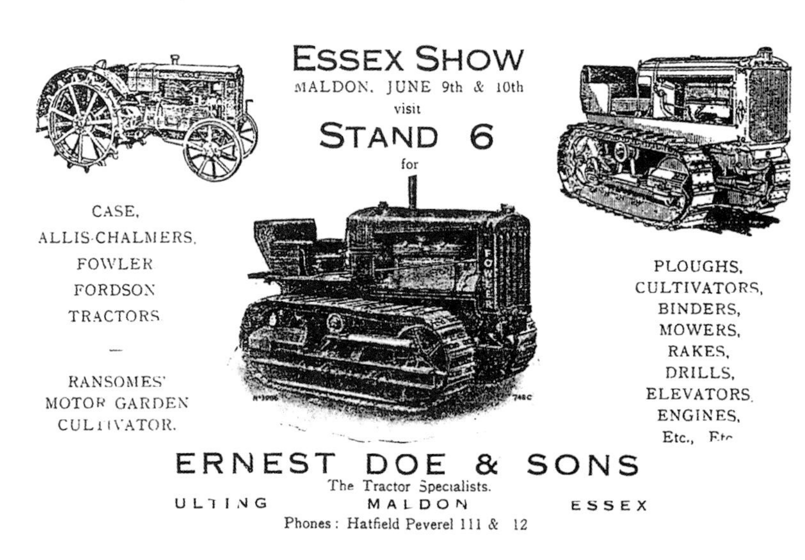

An advertisement for Ernest Doe & Sons' stand at the 1937 Essex Show. It serves to demonstrate the range of tractors and equipment handled by the firm at the time.

Doe's Corner photographed from a field of Onward peas in April 1939. The original blacksmith's shop can be seen on the right of the photograph, while the house on the left, Hill View, was built by Ernest Doe in 1906. The hanger in the centre formed the new workshop and had been acquired from the Eastern National bus company in Chelmsford after it became redundant when double-deckers were introduced.

and Allis-Chalmers tractors in 1930 and 1934 respectively. In 1937, Ernest Doe & Sons was formed with Ernest, Ernest Charles and Herbert as partners. The firm advertised itself as 'The Tractor Specialists' and held agencies for Case, Allis-Chalmers, Fowler, Fordson and Ransomes machines.

The responsibility for running the firm fell increasingly to Ernest Charles, who also managed the tractor sales. In addition to the repair and retail side of the business, the partnership also offered an agricultural and industrial contracting service. This side was run by Herbert Doe while the third brother Hugh farmed both on his own account and with other land owned by the Doe family.

A Case LA in Ernest Doe & Sons' yard in 1941. It was one of 240 new tractors sold by the firm that year. The old Peterbro, Alldays, IHC Mogul and Fiat tractors in the background were all scrapped as part of the war effort.

This Case machine was one of the first combines sold by Ernest Doe & Sons. The firm had held the Case agency since the 1920s, but by the Second World War it was also selling Allis-Chalmers, Fordson, Fowler, David Brown, Oliver , Caterpillar and Ransomes equipment. Ernest Charles Doe can just be seen to the right of the tractor driver.

The clamour for machinery to increase food production during the Second World War saw a further growth in Doe's business. In 1941, the firm sold a total of 240 new tractors - 94 Fordson, 8 David Brown, 62 Allis-Chalmers, 51 Case, 12 Caterpillar, 12 Cleveland Cletrac and 1 Oliver 80. The same year, Ernest Charles even allowed his collection of fifty veteran tractors to be scrapped for the war effort.

To save petrol during the war, it was suggested that Essex be divided up into sections with each agricultural dealership operating in its own zone. This would seriously have affected Doe's business and lost it many of the customers that it had outside its own area. To counter this, the firm opened branch depots at Fyfield and Colchester in 1943. The Fyfield branch began in buildings at Pickerells Farm that had belonged to the Doe family since 1939, while the Colchester premises was a former coal merchant's yard. A third branch at Gosfield was added in 1945 and was acquired from a Mr Farleigh who made trailers from old lorry axles.

Doe became Massey-Harris agents and added harvesting equipment to its already impressive stock list. The Massey-Harris agency was an important franchise for the firm as it gave it the opportunity to sell the new Model 21 self-propelled combines. After the war, Doe bought a number of ex-government Allis-Chalmers HD crawlers and reconditioned them for sale to farms across the country. The firm also replaced the petrol-paraffin engines in a number of ex-ministry Caterpillar R4 crawlers with diesel power units.

Ernest Doe & Sons' stand at the 1951 Royal Show. Left to right are a Fordson E27N equipped with a Dorman sprayer, a David Brown Super Cropmaster, a Nuffield Universal, a David Brown Trackmaster, a Hanomag crawler, an Allis-Chalmers Model B and a Tayler-Doe silage harvester.

A Massey-Harris 744D tractor inside Doe's main workshop in about 1953. This building, originally part of the bus depot in Chelmsford, still forms part of the complex of buildings at Ulting today.

The stores at Ulting during the 1950s. Ernest Doe & Sons Ltd. had been incorporated in 1947 and Massey-Harris remained one of its most important franchises until it became a Fordson main dealer.

The postwar boom in sales and the resulting increased turnover led to the decision to turn the business into a limited company. Ernest Doe & Sons Ltd was incorporated on 15 March 1947, with Ernest Charles as chairman and managing director. Joining him on the board were his son, Alan, and brother, Herbert. The company opened further depots, and legend has it that Ernest Charles bought three branches, Braintree, Rochford and Stansted, on one Saturday morning in July 1947. The same year, the Colchester depot was formed into a separate business, Colchester Tractors, in order to get the Fordson franchise for the area. Four years later, Doe moved into Suffolk, opening a branch at Sudbury in 1951.

By the 1950s, Ernest Doe & Sons was selling more Fordson tractors than any other make and was under pressure to become exclusively Ford dealers and give up the competing franchises. The alternative was to sign up as a Massey-Harris-Ferguson dealer. In the end, the company felt that the Fordson tractor was too strong a product to lose, and was appointed a Ford main dealer with effect from 1 February 1956.

The bond between Ernest Doe & Sons and the Ford Motor Company grew into a strong alliance. Before long,

Ernest Charles Doe, the eldest son of the founder and the first chairman and managing director of Ernest Doe & Sons Ltd from 1947 to 1971. From a lad, he had been keenly interested in tractors and was instrumental in securing most of the company's important franchises and the introduction of the Triple D.

Ernest Doe senior (left) pictured in the blacksmith's shop at Ulting in 1958. This is now the site of the sales office. Ernest Doe sadly passed away in 1964, aged eighty-eight years.

Fordson tractors arriving at Ulting on Ernest Doe & Sons' Ford Thames Trader lorries in about 1960, possibly returning from a show. The company was appointed a Ford main dealer in 1956, and before long was selling around 500 Fordson tractors a year.

The Braintree depot was one of three branches bought by Ernest Charles Doe on one Saturday morning in July 1947. This photograph, taken in about 1958, shows Fordson tractors lined up outside the stores. A Roadless Power Major can be seen in the centre with an industrial Dexta to its right.

Doe was selling around 500 Fordson tractors a year, and expanded its sales territory across the south-east of England. The company also entered the export market, sending a hundred Fordson E27N Majors a month to Australia and New Zealand in conjunction with J. J. Wright & Sons Ltd of Dereham in Norfolk.

The construction machinery side of the business began when the company started selling Weatherill, Bray and Whitlock loading shovel and digger conversions of Fordson tractors before taking on the JCB franchise in 1959. After JCB dropped Ford skid units in favour of BMC-based machines in 1965, Doe changed over to Ford's own industrial equipment range.

The company also inaugurated its own show in February 1960, inviting customers from across East Anglia to view equipment at the Braintree branch. The

*Two Super Majors and a Dexta tractor in the snow outside the Stansted depot in early 1961.
The depot on Lower Street in the centre of the town was one of the three branches acquired in 1947.*

The Sudbury depot on Cornard Road was opened in 1951. It was Ernest Doe & Sons' first branch in Suffolk. The German Kemper manure spreader on the right of the forecourt was imported through the associated Colchester Tillage company.

A new JCB2 excavator/loader leaves the yard at Ulting in 1963. Ernest Doe & Sons held the JCB franchise from 1959 until 1965 when it was dropped in favour of the Ford industrial range. The lorry is a Mark 1 Ford Thames Trader that was built in 1960. A Triple D tractor can just be seen behind the low-loader.

'Doe Show' was a great success and became an annual fixture. It was moved to Ulting in 1962 and expanded to include a comprehensive working demonstration of tractors and implements on land adjacent to the works. The event was soon attracting visitors from all over Britain.

Ernest Doe senior passed away on 1 October 1964, aged eighty-eight years. The same year, his grandson, Alan, was appointed joint managing director, joining his father, Ernest Charles, at the head of the company.

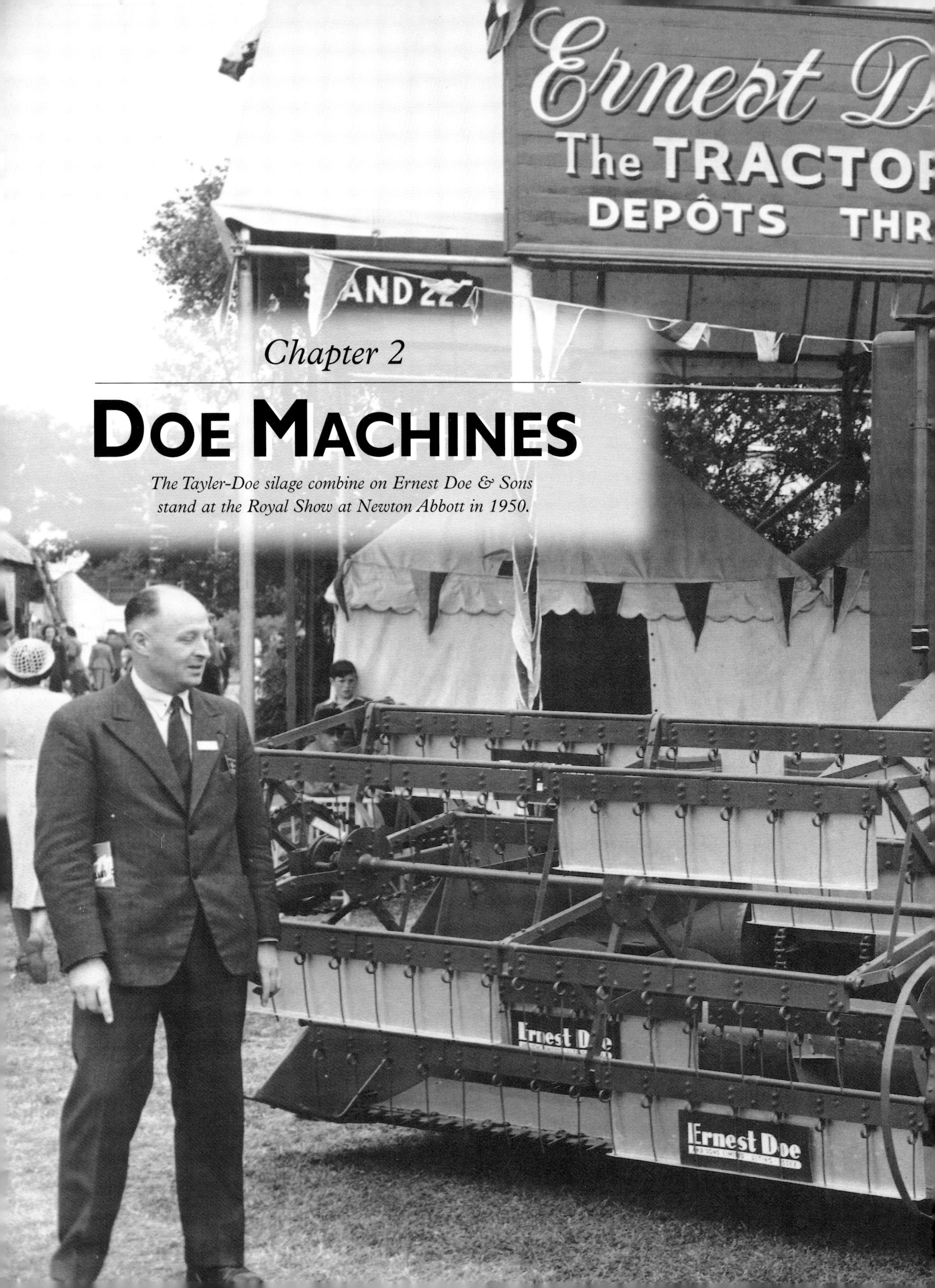

Chapter 2

DOE MACHINES

The Tayler-Doe silage combine on Ernest Doe & Sons stand at the Royal Show at Newton Abbott in 1950.

& Sons Ltd
SPECIALISTS
HOUT ESSEX
DASH
STAND 226
SILAGE
Ernest Doe
SILVER MEDAL

Although Ernest Doe & Sons became better known as a retail organisation, it must not be forgotten that the company was born out of the blacksmith's trade and was prepared to repair and manufacture where necessary. Furthermore, Ernest Doe senior had something of a reputation as an inventor and had several interesting projects on the boil during the 1930s.

One of Ernest's first inventions was a rotary trencher based on a Case Model L tractor. It was evidently a crude affair, but worked reasonably well and was used to drain a grass airfield for the Air Ministry by the contracting arm of Ernest Doe & Sons. Another project was a mechanical system for loading stooks of corn from the field. It consisted of a binder platform and sails connected to an elevator. Dubbed the Hurdy-Gurdy, it was only moderately successful and was not put into production.

The 1940s saw Ernest Doe working on a mechanical potato planter. The design used a reciprocating floor to metre the seed potatoes. Although only one prototype was built, it is believed that Transplanters Ltd of Sandridge near St. Albans considered using the machine as the basis for its Robot potato planter following liaison between the two companies. The first machine to be actually marketed as a Doe product is believed to have been a conversion of the Lister Auto-truck that was turned into a self-propelled sprayer for horticultural use.

In order to deal with urgent repair jobs in the field, Doe converted a Fordson into a mobile welding outfit in 1943. The tractor, an early 'water-washer' Model N on pneumatic tyres, had a Murex electric welder mounted on a frame over the fuel tank. The welder was a self-contained unit with its own generator that was driven by multiple vee-belts from an adapted pulley on the tractor. A workbench incorporating a toolbox to hold the electrodes and welding equipment was built into a framework on the side of the tractor opposite to the pulley drive. The company put the adaptation on the market, and later conversions based on the Fordson Diesel Major featured a rear-mounted weighing unit.

While holding the Allis-Chalmers agency, Ernest Doe & Sons converted a number of Model B tractors to low

A 1943 Doe conversion of a Fordson tractor into a mobile electric welder for repairs in the field. The Murex welder was mounted over the fuel tank and driven by a vee-belt taken off the tractor's pulley drive.

Doe's workshop manager, Charles Bennett, at the wheel of a later Doe-Murex conversion of a Fordson Diesel Major. The welder was mounted at the rear and driven off the belt pulley attachment. The machine was sold to a firm in London that used it for erecting steel buildings.

A Doe orchard conversion of the Allis-Chalmers Model B tractor. The tractor was converted to low clearance by rotating the drop housings through 90 degrees and modifying the front axle.

A Doe high-clearance conversion of the Allis-Chalmers Model B straddles one of the company's orchard tractors. Note that the low-clearance machine is fitted with streamlined shields for orchard work.

clearance by rotating the drop housings through 90 degrees and modifying the front axle arrangement. The company also offered streamlined shields to protect the tractor and operator for orchard work. Similar low-clearance conversions were produced by Drake and Fletcher of Ashford in Kent on the later Allis-Chalmers D272 during the 1950s.

Doe was also responsible for a special high-clearance version of the Allis-Chalmers Model B that was made by mounting the tractor on stilts with an enclosed chain-drive to the rear wheels. The first of these machines was ordered by Seabrooks, a local fruit grower, and was used for spraying blackcurrant bushes with DNC-petroleum or lime-sulphur.

In 1948, the company demonstrated a remarkable gadget called the Doe Wild Oat and Weed Seed Collector. This ungainly contraption, known around the Ulting works as the 'V1' after Germany's rocket weapons of the Second World War, was designed to collect wild oats, white clover and weed seeds off the ground after the combine. It was built by Doe for one of its customers, A. H. Gardiner, who co-operated in the design.

The machine was basically a glorified vacuum cleaner mounted on a Fordson E27N Major tractor. A Keith Blackman fan, driven off the Fordson's belt pulley and running at 2,500 rpm, provided the suction. The collector at the front used a series of spring tines to agitate the surface of the ground and loosen the seeds, which were then drawn up a large tube to be blown by a blast of air into a covered trailer towed behind the tractor. The arrangement was not a great success as the E27N's engine had hardly enough power to provide adequate suction for the system to work.

The Tayler-Doe silage combine was a self-propelled green-crop harvester, built for cutting and loading grass

The Doe high-clearance tractor was mounted on stilts and had an enclosed chain-drive to the rear wheels. It was designed for spraying blackcurrant bushes, and the first one was sold to the local growers, Seabrooks.

for drying. It was the brainchild of R. G. Tayler, the area manager for Strutt & Parker (Farms) Ltd one of Doe's larger customers. Strutt & Parker's Burnham-on-Crouch farm grew 350 acres of grass mixtures and lucerne for conversion into meal by a Templewood drying plant. The silage combine was designed to speed up the operation and replace no less than five tractors and mowers, together with a number of swath turners and crop collectors.

Introduced in 1948, the Doe Wild Oat and Weed Seed Collector was a remarkable contraption designed to collect wild oat, white clover and weed seeds off the ground behind the combine. It was basically a glorified vacuum cleaner with the suction provided by the belt-driven fan that can be seen mounted on the front of the tractor.

The Tayler-Doe silage combine was a self-propelled green-crop harvester built in conjunction with R. G. Tayler of Strutt & Parker Farms. Seen in action in 1950, it was capable of cutting and loading 3 tons of green material in less than ten minutes.

The Tayler-Doe silage combine on the company's stand at the 1952 Royal Show at Newton Abbot where it was awarded a silver medal and the Burke Trophy. With the machine (left to right) are Alan Doe, Ernest Charles Doe, R. G. Tayler and Lionel Harper of Massey-Harris.

The Domobile was a cable-operated universal excavator based on an ABS machine imported from Sweden. Ernest Doe & Sons fitted the excavator with a Fordson Major diesel engine mated to a Ford Thames truck gearbox. It was capable of 25 mph on the road.

The machine was based on a Massey-Harris 726 combine harvester with the necessary adaptations carried out in the workshops at Ulting. All the threshing and grain collection parts of the combine were removed, leaving a basic chassis with the engine and drive system intact. The header was retained and fitted with a Reynolds pick-up reel and a serrated knife to cope with the increased moisture of the green material. The crop was fed from the table auger to a stripper-beater fitted with a knife attachment where it was chopped before being elevated into a trailer towed behind the machine. The speed of the combine's operating mechanism was

Launched in 1957, the Doe-Dorman was a 200 gallon self-propelled sprayer based on a Fordson Major tractor. It featured saddle tanks, sprung folding booms and an air-conditioned cab.

also increased to match the workload.

The prototype silage combine went to work on Strutt & Parker's farm during the 1950 season and proved itself capable of cutting, elevating and loading three tons of green material in less than ten minutes. With Massey-Harris' approval, Doe put the machine into production and began taking orders for the 1951 season. The customer could buy either a conversion kit that cost £250 or a complete machine for £1,250. Several Tayler-Doe silage combines were sold in Essex as well as throughout the West Country, and the machine was awarded the Burke Trophy and a silver medal at the 1952 Royal Show, held at Newton Abbott in Devon.

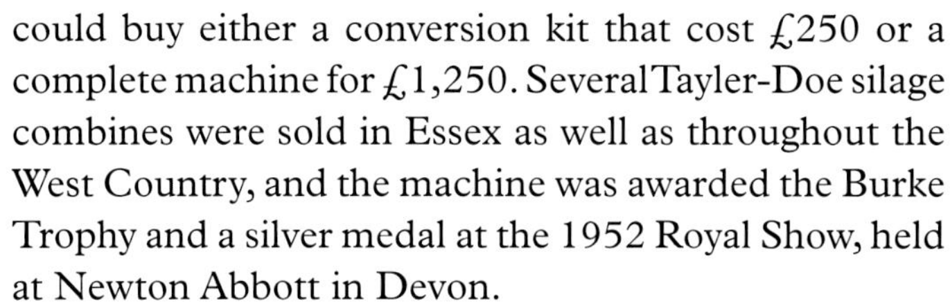

Another venture for the company at this time was a cable-operated universal excavator known as the Domobile. It was based on an ABS machine imported from Orebro in Sweden. Its original engine, believed to have been a Bolinder-Munktells power unit, was replaced by a Fordson Major diesel engine mated to a Ford Thames lorry gearbox. The conversion, along with further modifications to meet UK regulations, was carried out at Ulting.

The Domobile was a wheeled machine with pneumatic tyres that was capable of travelling at 25 mph on the road. It weighed 6½ tons and could be used as a dredger, grab, crane or face shovel. It was priced at £3,450. Very little interest was shown in the excavator once the more successful Whitlock and JCB machines appeared on the market.

The Doe-Ransomes fork-lift attachment was demonstrated at the 1959 International Potato Harvesting Demonstration. Designed to fit the Fordson Dexta, it had a 15-cwt capacity with a mast assembly taken from a Ransomes electric fork-lift truck.

The Walker-Doe ditcher for excavating and maintaining drainage ditches was launched at the National Hedge Laying and Ditching Contest, organised by *Farmers Weekly* and held at Leamington Spa in January 1959. The machine consisted of a Whitlock Dinkum Digger Major hydraulic excavator attachment mounted on a Track-Marshall 55 crawler. The complete unit, including the tractor fitted with a hydraulic angledozer, was priced at £3,989. A specially shaped ditcher bucket for drainage work cost £85 extra. The rear-mounted excavator was replaced by a JCB Hydra-Digga backhoe in November, but Doe only marketed the machine for a short time. The idea was eventually taken over by Brown Brothers of Edinburgh, who built a modified version known as the Brown BXT55.

A Fordson Super Major with a New Holland 818 precision-cut forage harvester. The harvester has been fitted with a six-cylinder Ford 590E auxiliary engine by Ernest Doe & Sons to increase its output when chopping grass for drying purposes. Note the dual wheels to support the weight of the engine.

The Doe-Dorman high/low-volume sprayer was based on a Fordson Major fitted with 200-gallon saddle-tanks. The machine was ahead of its time in that it featured an air-conditioned cab, pressurised by a pto-driven fan. Built in conjunction with the Dorman Sprayer Company of Cambridge, the unit had 30 ft folding booms and was launched at the 1956 Smithfield Show.

The International Potato Harvesting Demonstration, held near Southend in October 1959, saw the unveiling of the Doe-Ransomes fork-lift attachment for the Fordson Dexta tractor. The mast assembly was taken from a Ransomes battery-powered fork-lift truck, and the unit would lift 15 cwt. A 25 cwt version for the Fordson Major followed later.

The workshops at Doe's Corner were never idle, continually adapting machinery for agricultural or industrial work. May 1963 saw the company fit a few New Holland 818 precision-cut forage harvesters with Ford six-cylinder 590E diesel engines. The conversion, which included dual wheels to support the weight of the engine, was carried out to increase the output of the forager when chopping grass as short as half an inch for drying purposes.

One of the company's most interesting projects was a road roller based on a Fordson Major tractor. The conversion was carried out at the Braintree depot and the machine was fitted with a hydrostatic forward/reverse shuttle. Only one example was built and it was hired to Cambridge University when colleges were being extended. It ended its working life on a caravan site before being bought for preservation and restored by collector and farmer, Andrew Streeter, of Bishop's Stortford. He has since sold the roller and it is now earning its keep rolling lawns on an estate near the south coast. The front roller on this unique machine was supported by a massive sub-frame and the steering system used several components from the company's tandem tractors - the famous Triple Ds that became the best known of all Doe's products.

The Doe road roller in 'as found' condition prior to restoration by collector Andrew Streeter. Note the substantial sub-frame to support the front roller. The forward/reverse shuttle lever to control the hydrostatic transmission can be seen just to the right of the fuel tank.

The Doe road roller behind the fence at the Ulting yard in the 1950s. Note its original cab.
The large works building in the centre of the photograph became the main assembly shop for the Doe Triple D tractors.

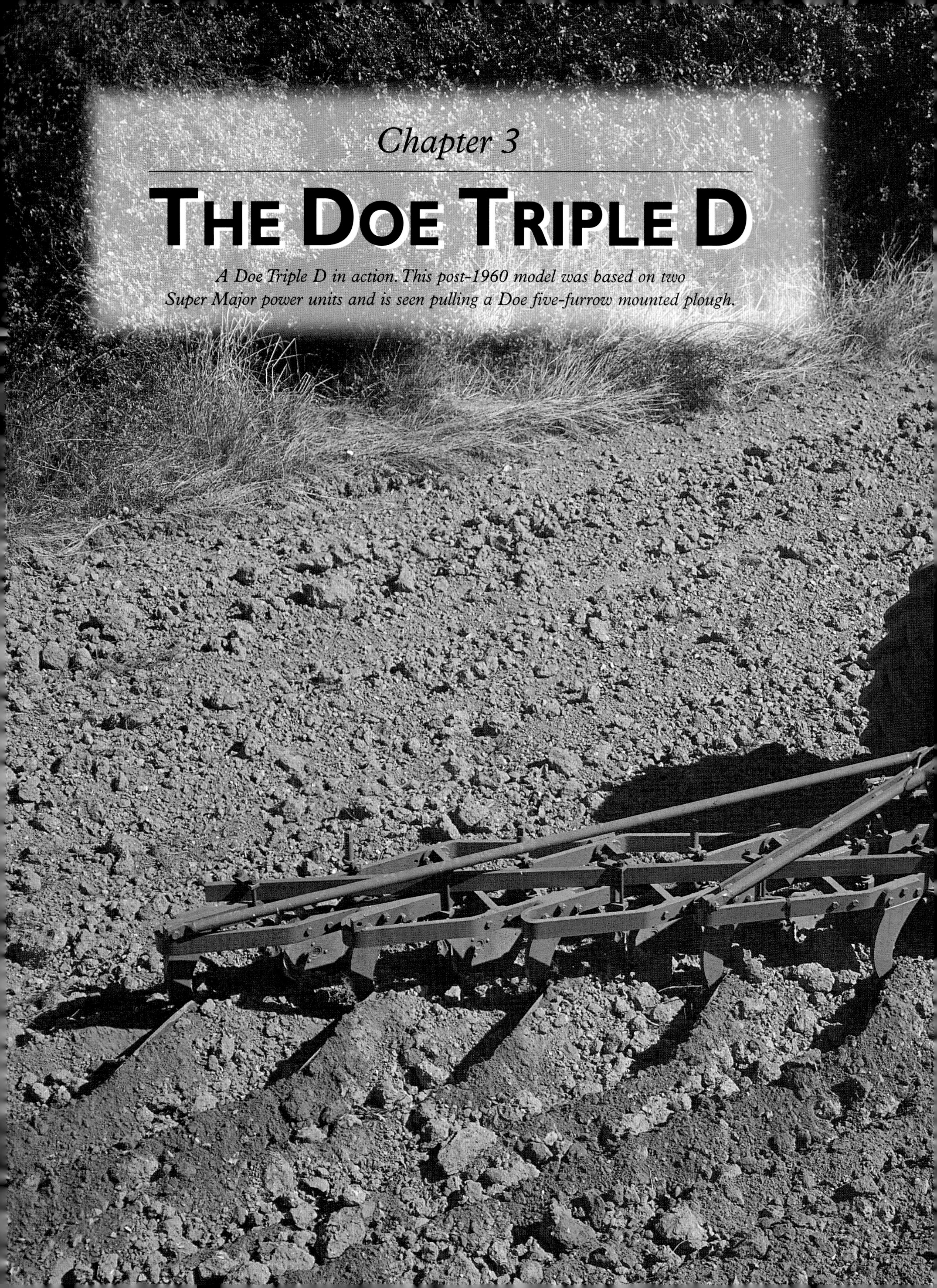

Chapter 3

THE DOE TRIPLE D

A Doe Triple D in action. This post-1960 model was based on two Super Major power units and is seen pulling a Doe five-furrow mounted plough.

SUPER MAJOR

The Doe Triple D must rate as one of the world's most unorthodox tractors. It was a remarkable machine with four driving wheels, two power units and articulated steering. However, its ungainly appearance was deceptive, and its level of performance was astonishing for the time.

The Triple D story began in 1957, when a local farmer, George Pryor of Navestock, was looking for a machine to replace his County crawler and cope with his large acreage of heavy Essex clay - land which, in his words, 'was as tough as you could find anywhere'.

Crawlers did not suit Pryor's farming system; he believed them to be too slow and restrictive. He wanted more power, greater versatility and the capability to handle large and heavy cultivation equipment. A 100 hp wheeled tractor with plenty of traction would be ideal but there was nothing suitable on the market. Undeterred, he decided to develop his own machine in the farm workshop, basing his concept on the idea of two tractors working in tandem.

George Pryor already had something of a reputation as an inventor. He grew a lot of potatoes and had perfected a box-tippler mechanism for handling wooden potato boxes. It was fitted to a Ferguson high-lift loader in June 1955 and was successful enough for him to consider putting it on the market. Building his own tractor was just another project.

Preliminary experiments began in the autumn of 1957 and led to two Fordson Diesel Major tractors being joined together by a beam. A driver was still needed on each tractor, and the leading machine still retained its front axle. While the combination worked, it was far from satisfactory. It had loads of power but not enough traction. Calculations proved that 20 per cent of the tractive effort was being wasted in just pushing the front axle along.

The next stage of the development was to ditch the front axles and use a turntable arrangement to join the tractors together and provide a pivot between the two units. The turntable was based on a third-wheel coupling taken from an articulated lorry. The two tractors were steered by a single ram connected to a chain that operated on the turntable. Removing the front axles had a dramatic effect on the machine's performance, which was further enhanced by fitting new tyres to the two tractor units, but the steering arrangement still left much to be desired.

Encouraged by what he had achieved so far, George Pryor returned to his workshop and began fabricating a new turntable and a new sub-frame. The support mounting for the front tractor unit was made from a steel channel-section beam secured to the drawbar anchorage. The beam extended from the rear of the tractor and was supported by angled gusset plates bolted to the back of

The Doe's Dual Power demonstration tractor at Boyton Hall in late 1958. This was the second prototype, built up by George Pryor but fitted with Fordson Power Major skid units at Ulting.

the transmission housing. Machined steel tube was welded to the beam to form a trunnion extension. The turntable was mounted on a sub-frame assembly under the front of the rear tractor unit. A steel tube welded to the underside of the turntable was fitted with phosphor-bronze bushes to accommodate the trunnion extension from the front unit and form a pivot shaft to allow lateral movement between the two tractors.

The difficulty with the pivot-steer design was to get the steering rams to 'go round the corner' as the tractor articulated. Pryor's solution was to use a floating centre plate in the turntable to which four rams were attached. The hydraulic rams were arranged in two pairs of two, working in tandem either side of the turntable. The four rams gave a smooth steering action while the centre plate acted as a joint between the rams and allowed them to articulate with the tractor. George Pryor patented the design during 1958 and it was filed under British Patent No. 6777/58.

The steel turntable had a diameter of 2 ft and its kingpin was carried on heavy-duty roller bearings. Proprietary hydraulic loader rams were used as steering rams and were controlled from a valve chest mounted on top of the rear tractor's transmission housing. The steering wheel was dispensed with, and a lever connected to the valve chest was used to tiller-steer the machine.

Fordson Majors were chosen for the design because they were uncomplicated, rugged and reliable, and were one of the few tractors with suitable clearance under the sump to allow for the conversion. George Pryor had also found a cheap way of sourcing suitable Fordson skid units; he went to the Cambridge Machinery Sales and bought good second-hand examples of the vaporising oil models. These were not as popular as the Diesel Majors, but had often done very little work and could be bought for a fraction of the price. He then replaced the vaporising oil engines with new diesel power units obtained from Rotary Hoes of West Horndon. Rotary Hoes was buying in new Major skid units for its Howard trenchers, fitting them with Perkins L4 engines for more power and then selling off the unused Fordson diesels for around £120 to £130 each. This arrangement meant that Pryor ended up with virtually new skid units for far less than it would have cost him to buy new tractors for the conversion.

Field trials during 1958 proved that George Pryor's tandem tractor was a success. It looked complicated, but the resulting tractive ability of the arrangement went beyond all expectations. The combination of the two 52 hp Fordson power units provided a four-wheel drive machine with over 100 hp available. It pulled twice as

The Doe's Dual Power is demonstrated at Boyton Hall on land belonging to Hugh Doe. The field had recently been cleared of sugar beet and was in a very poor state, but the tractor coped admirably with the sticky conditions. George Pryor's driver is at the wheel.

An early Doe Dual Power at work, again with George Pryor's driver at the wheel. This was probably Pryor's own tractor, based on his original prototype with a few modifications.

much as the old County crawler, and at more than twice the speed.

The tractor created great interest and became the talk of the area. Pryor, one of Doe's long-standing customers, was both a friend and shooting pal of Ernest Charles Doe. During one particular pheasant shoot, the topic of the tandem tractor came up in conversation. Ernest Charles asked if he could bring Carlton Whitlock of the excavator and trailer manufacturers, Whitlock Brothers of Great Yeldham, to appraise the machine's potential. Whitlock was not convinced that the tandem tractor design had any future, but Ernest Charles was keen to see the machine in production and decided to secure the manufacturing rights for his own company.

An agreement was reached with George Pryor who then built up the framework for another tractor. This second prototype was sent to Ulting where it was fitted with Fordson Power Major skid units. Doe's works manager, Charles Bennett, was put in charge of the project and was largely responsible for adapting Pryor's ideas for production. He instigated a few changes and replaced the tiller with a steering wheel. Rather than use the loader rams, the company had special steering rams made up at Rayne foundry in Braintree. The machine was ready for trials within three weeks and became Doe's demonstration tractor.

At over 20 ft long, the machine appeared cumbersome and ungainly, but was surprisingly manoeuvrable as the turntable allowed the two units to be swivelled almost at right angles. Amazingly, the turning circle at 21 ft was actually 5 ft less than that of the ordinary Fordson Major. Oil to power the steering arrangement was tapped from the standard hydraulic pump in the front tractor unit. This would have been otherwise redundant as the rear unit operated the hydraulic lift and controlled the implements. The idea behind this arrangement was to spread the load over the two engines as equally as possible.

The clutch operation on the two units was synchronised through a single pedal on the rear tractor using a hydraulic master cylinder in conjunction with a slave ram for remote control. The throttles were connected together by a wire.

The only control the driver had over gear selection on the forward tractor was through a system of link-rods running across the rear tractor bonnet, allowing him to

Ernest Charles Doe (left) and George Pryor (centre left) attend a demonstration of a later Triple D tractor. The prospective customer on the right of the group is farmer Jack Joice, the brother of the famous television presenter, Dick Joice. Next to him is his tractor driver who appears not to have dressed for the occasion.

operate the high/low gear lever on the front unit. With this lever in neutral, the driver walked round to the front unit to manually select the gear ratio required for the job in hand. On returning to his seat, he selected the same ratio with the main gear lever on the rear unit, slipped the front tractor out of neutral using the high/low linkage, and away he went. Changing gear could be a time-consuming task. Reversing on headlands was achieved by placing the high/low gear lever of the front unit in neutral and using the reverse gear of the back tractor only. It was not an altogether satisfactory arrangement, but it worked.

The Doe tractor had nearly perfect weight distribution and coped admirably with all weathers. Optimum results were obtained by setting the forward engine to run at a marginally higher speed than the back unit so that the front tractor 'pulled' the rear unit. Ploughing rates of two acres per hour were recorded, and the tractor could cultivate over seven acres per hour with a 20 ft spring-tine.

The tractor bore the legend 'Doe's Dual Power' on the bonnet, possibly to reflect the combination of two Power Major tractors. The first demonstration was held in the autumn of 1958 at Boyton Hall, near Roxwell, on land farmed by Hugh Doe. The demonstration followed a period of wet weather and the field, recently cleared of sugar beet, was in a very poor state. It was a good test of the Dual Power's capabilities and its performance was undeniably impressive.

George Pryor's own original tractor ran for ten years without problems before being dismantled. He and his brother subsequently bought a new Triple D each from Doe in October 1961. George also bought a new 130 in 1968, and later acquired a second-hand Triple D that he still owns and is destined for restoration. His agreement with Doe meant that he received a royalty payment for

Ernest Charles Doe (centre) and his works manager, Charles Bennett (right, with pipe), inspect a Fordson Power Major skid unit in the assembly shop at Ulting. Bennett was largely responsible for adapting George Pryor's ideas for production and this unit would become the front half of a Doe tandem tractor. Note the centre plates of the turntable that can be seen in the foreground.

each tractor sold. He admits that he did not make much money out of the arrangement as it just about covered the cost of taking out the original patent rights, as well as maintaining them and extending them to cover other countries.

The first Dual Power to be sold was delivered to Lordsfields Farms in west Suffolk on 22 October 1958. It proved to be a saviour during that year's wet autumn, and was the only machine able to cope with the abysmal conditions and continue ploughing after lesser tractors on the farm had fallen by the wayside. A second machine was also supplied to Suffolk in November, while the third Doe tractor, sold in December, stayed in Essex.

Two more of this early version of the tandem tractor were bought by farms in Essex during February 1959, making a total of six including the demonstration model that the company had retained for its own use. These proved to be the first and last to be badged as Doe Dual Power tractors, and there followed a gap of several months before the articulated tractor appeared again with a new name, a new colour scheme and, most importantly, a new steering system.

The early steering system had not been very successful. The drop arm from the steering pump connected directly to the Dowty control valve and only allowed the steering wheel about 2 in. movement. Just the slightest touch was needed to open or close the valve, resulting in very erratic steering and at least one driver putting his machine in the ditch. A return to the original tiller or a joystick steering arrangement was considered but was sensibly dismissed. Also, with the steering system being tapped out of the front tractor's hydraulic system, it did not work when the clutch was depressed.

The first Doe Dual Power to be sold was delivered to Lordsfields Farms in West Suffolk in October 1958. Only two more went out that year.

Ironically, the changes to the steering system were prompted not so much by its failings as by several Doe owners' need to tax their machines for road use. The problem was that the Dual Power tractors fell foul of the traffic regulations. It seems that the Motor Taxation Department had decreed that for the tandem machine to be classified and taxed as one vehicle, all the auxiliaries, such as the hydraulic system and power steering pump, had to be mounted on the rear unit and driven off the back engine. In order to satisfy the taxation rulings, a hydraulic pump to power the steering system was fitted to the rear tractor unit. Supplied by Plessey, it was a vane-type pump that was bolted to the front casting and driven off the end of the crankshaft through a duplex chain coupling.

At the same time, the opportunity was taken to improve

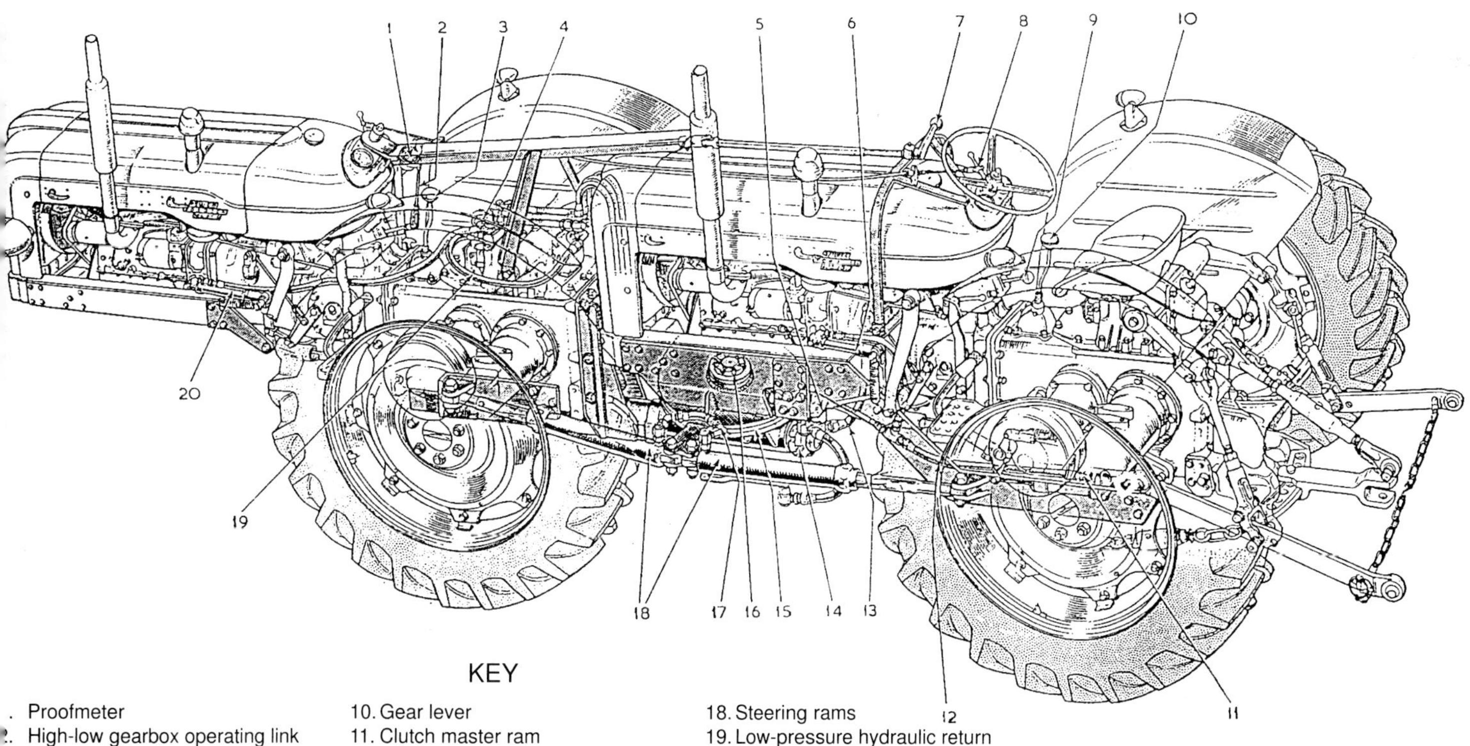

KEY

1. Proofmeter
2. High-low gearbox operating link
3. Gear lever
4. Steering hydraulic feed system
5. Steering control valve
6. Clutch slave ram
7. High-low operating lever (forward tractor)
8. Governor
9. High-low lever (rear tractor)
10. Gear lever
11. Clutch master ram
12. Hydraulic feed to clutch slave rams
13. Hydraulic feed to right-hand steering rams
14. Lateral pivots
15. Turn-table
16. Turn-table adjusting nut
17. Ram oil transfer pipe
18. Steering rams
19. Low-pressure hydraulic return
20. Clutch operating ram

An exploded view of the Doe Dual Power tractor. Note the rod-linkage gear-change and the oil supply for the power steering that was tapped from the front tractor's hydraulic system. (Farmer & Stockbreeder copyright)

A Dual Power tractor fitted with large-diameter Wright steel wheels handles a Doe seven-tine cultivator. Only six of these early tractors were built as the steering system fell foul of the motor taxation laws.

the steering system by fitting a Vickers control valve with a compensating ram between it and the linkage. The ram feathered the action of the control valve and made the steering more progressive, allowing the wheel to move through one and a half turns from lock to lock. Most of the early Dual Power tractors had their steering systems modified to the new arrangement, but at least one still has its old system intact and has been preserved by Norfolk collector David Hills.

The reappearance of the tandem tractor at the Essex Show in June 1959 marked the official launch of the machine into the marketplace. It had the improved steering system, sported distinctive orange bonnets and was called Doe's Dual Drive. Brochures were issued, press releases were sent out to the farming press and the tractor was demonstrated to the public.

This photograph clearly shows the oil supply for the power steering system on the early Dual Power tractors. Note the bar attached to the top of the lift housing that allowed the two steel feed-pipes to move as the tractor articulated. The small copper pipes were the hydraulic feed for the clutch slave ram.

The Doe's Dual Drive, or Triple D, is launched at the Essex Show in June 1959. The tractor has drawn a large crowd and can only just be seen in the background on Ernest Doe & Sons' stand.

The Triple D had arrived, and the first was sold on 1 August 1959. The tractor caused plenty of interest, but many farmers still regarded it as a novelty and only another ten units left the works that year. However, 1959 saw the company's first export successes with two going to Russia through Avtoexport and another to Sweden. Three more units were supplied between January and March 1960 before production of the early model Doe tractors came to an end after a total of twenty had been built. The last machine with the rod-linkage gear-change was sold on 15 March to Boyton Hall Farms to work on 800 acres.

The introduction of the Triple D in 1959 marked the official launch of Doe's tandem tractor into the market place and brochures were issued for the first time.

The Triple D featured an improved steering system with a compensating ram (as seen in the centre of the photograph, above the large steering ram and behind the turntable) to feather the action between the linkage and the valve. It made the steering more progressive and the tractor easier to control.

Production of the Triple D was temporarily suspended for six months because Health and Safety officials had become a little concerned about the method of selecting neutral on the front unit by use of the rod arrangement. They believed it could be unsafe, particularly if the linkage became worn or sloppy. Because there was no detent on the neutral position on the high/low lever, it was difficult for the

Sand has been laid down to demonstrate the tight turning circle and manoeuvrability of the Triple D on Doe's stand at the Essex Show. The tractor could articulate through nearly 90 degrees and had a turning circle of 21 ft.

An early Power Major Triple D with the rod-linkage gear-change. Most Triple Ds were finished in Fordson Empire blue with distinctive orange bonnets, but the yellow and orange paint scheme on this tractor may signify that it was used for industrial work. Note the heavy-duty 14 x 30 wheels and tyres.

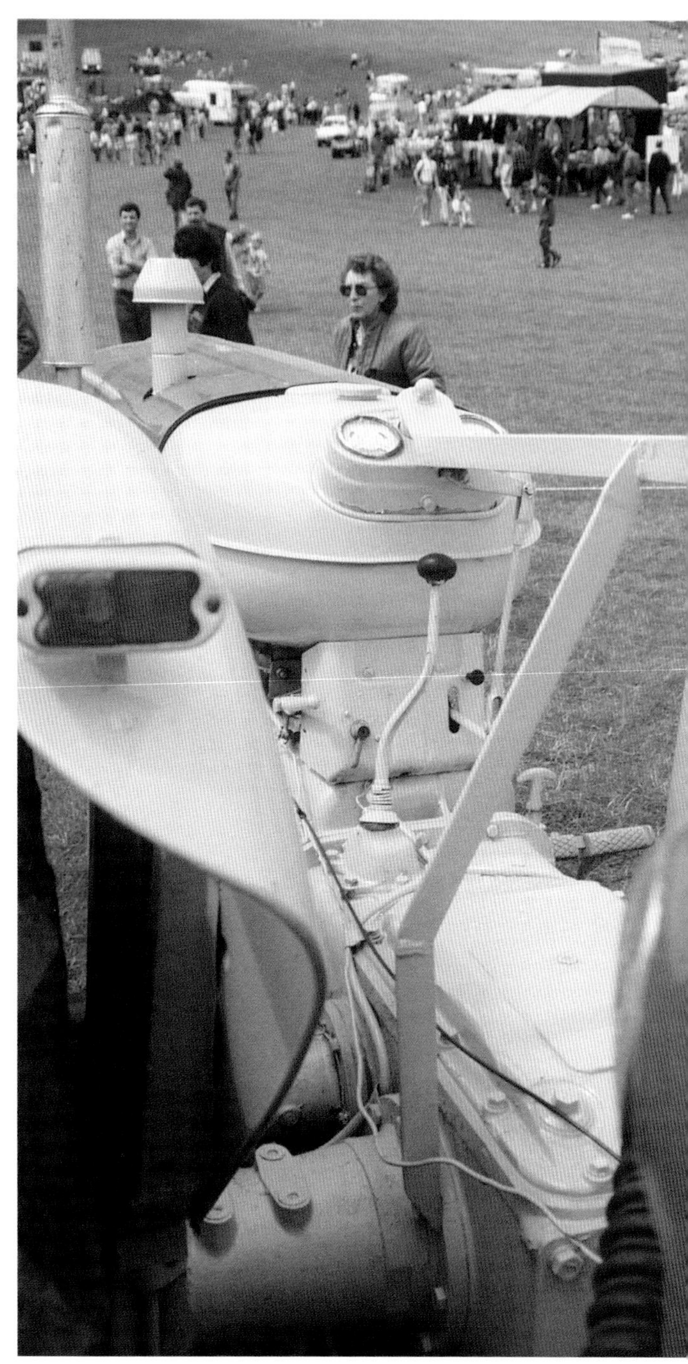

A close-up of the rod-linkage arrangement used to operate the high/low gear lever on the front unit of the Doe. Only twenty Dual Power and early Triple D tractors used this gear-change system before it was replaced because Health and Safety officials believed it unsafe.

operator to be absolutely certain when the front tractor was out of gear. A number of modifications were carried out at Ulting and selection of the high/low gear on the front unit was improved by using a hydraulic master cylinder and slave assembly in place of the link-rod arrangement.

A further development saw the introduction of another gear-change mechanism that allowed the driver to select all the gears on the front unit from the back tractor. It consisted of a system of master and slave cylinders that operated on the front gear selector, which had been shortened and was held captive by a fork in an enclosed box. A long lever on the right of the driver's seat was pushed forward to select first gear on the front tractor. Pulling it back engaged second. With the lever in a central (neutral) position, the gear selector on the front tractor could be

The Triple D was re-launched in May 1960 with an improved gear-change using a hydraulic master cylinder and slave assembly to operate the high/low lever in place of the rod-linkage arrangement. A full hydraulic gear-change mechanism was available, but was not fitted to this tractor because the primary gear lever can still be seen in place on the front unit.

This restored Power Major Triple D has the optional full hydraulic gear-change that allowed the driver to select all the gears on the front unit from the back tractor. The box containing the necessary slave rams can just be seen over the transmission on the leading tractor.

BELOW: The drawing details of the Doe Triple D version based on the Power Major. (Farm Mechanization copyright)

The improved Doe Triple D showing details of both the hydraulic high/low lever slave assembly and the revised steering system. (Farm Mechanisation copyright)

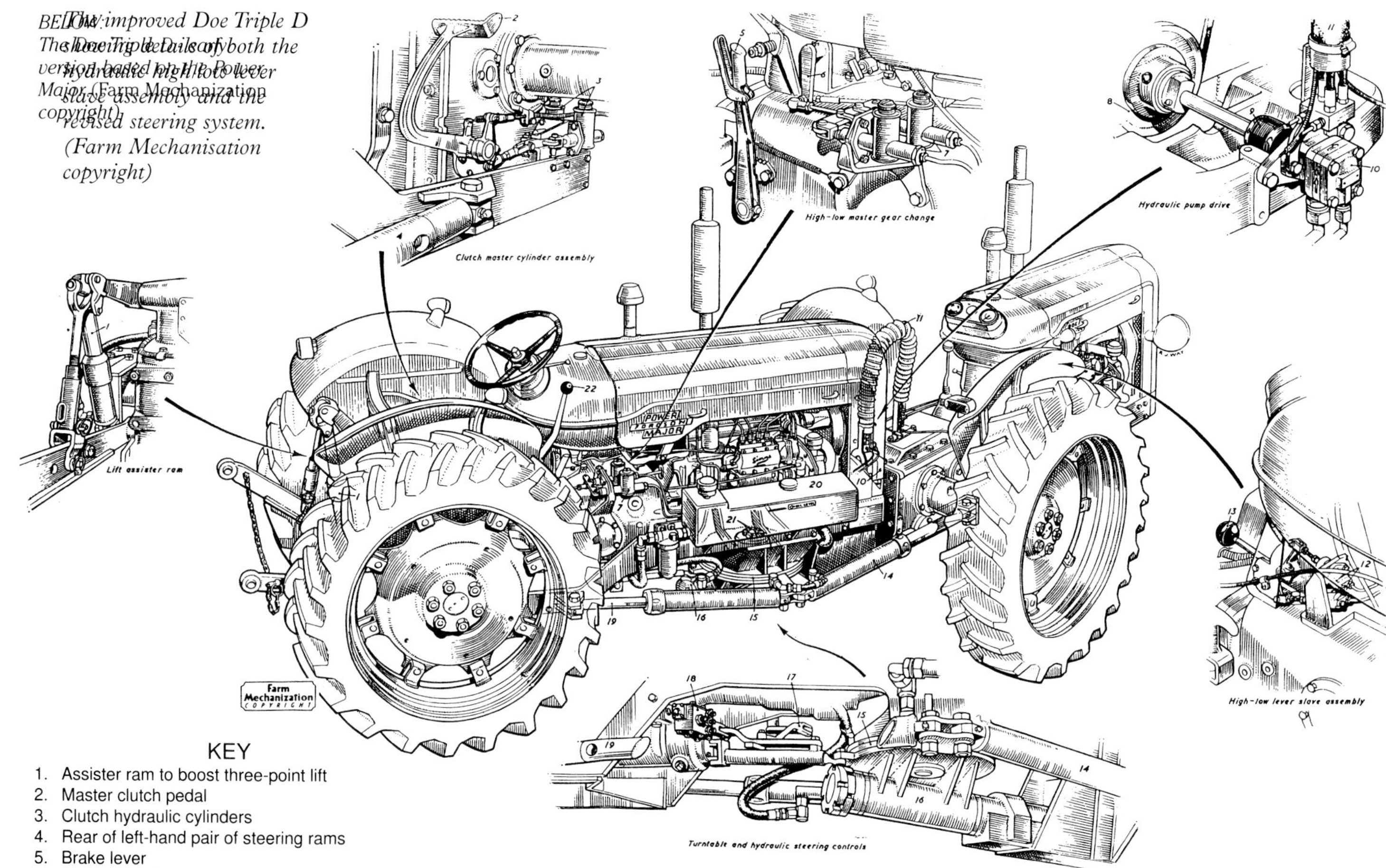

KEY

1. Assister ram to boost three-point lift
2. Master clutch pedal
3. Clutch hydraulic cylinders
4. Rear of left-hand pair of steering rams
5. Brake lever
6. Master high-low gear lever crank
7. High-low hydraulic cylinders
8. Fan belt pulley on rear engine
9. Hydraulic pump drive flexible coupling
10. Hydraulic pump for power steering
11. Conduit carrying clutch, high-low, throttle and pump cut-off controls to front unit
12. Hydraulic slave cylinders for remote control of front high-low lever
13. Front high-low lever
14. Front of right-hand pair of steering rams
15. Turntable assembly
16. Pivot shaft connecting front unit to turntable
17. Linkage between steering wheel and steering hydraulic control valve
18. Steering hydraulic control valve
19. Rear steering
20. Hydraulic steering oil-supply tank
21. Turntable pivot shaft
22. Throttle control lever coupled to both engines

121

moved across the gate by pressing a foot pedal on the rear unit. With the pedal depressed, the lever could then be used to select third or reverse gears. Once the lever was returned to neutral, the selector would automatically be pushed back by a spring to the first/second gear position.

It took a certain amount of work and experimentation at Ulting to get the full hydraulic gear-change right. The system was devised by Charles Bennett and fabricated by one of the fitters, Vic Spitty. It was originally offered as an option for an extra £75 but soon became standard specification on all the Triple Ds sold. The pipework for the hydraulic controls was enclosed in a flexible rubber hose that passed between the two power units.

The box containing the full gear-change mechanism as fitted to the later Triple D tractors. The lid has been removed to show the shortened primary gear lever held captive by a fork. The fore and aft slave rams operated on the fork to select the gears, while a single slave ram in the end of the box moved the lever across the gate. The large spring allowed the lever to return back once pressure was released. The unit is shown on a Super Major tractor.

The remote master cylinder arrangement for the gear-change on a Triple D tractor. The uppermost pair of cylinders operated the high/low gear lever. Below these are the two cylinders that activated gear selection on the front unit. The pedal above the foot brakes and to the right of the handbrake operated the lower cylinder that moved the primary gear lever across the gate. The power steering reservoir and filter assembly can be seen at bottom right.

Other modifications included replacing the wire linking the throttles with a long heavy-duty Bowden cable, which cost £100 and was one of the most expensive components on the tractor. An assister ram was added to the basic specification, and a solenoid-type starter for remote starting the front unit from the rear seat was another available option costing an extra £25.

The revised Triple D was re-launched in May 1960. Its basic price was £1,950 on 11 x 36 tyres. By comparison, an equivalent (93 hp) Caterpillar D6 crawler cost nearly £7,000, making the Doe a very competitively priced alternative to a tracklayer with far cheaper running costs.

A variety of wheel equipment, including 12 x 38 or 14 x 30 tyres, was available for the Triple D at extra cost.

All the hydraulic controls and the throttle cable for the front tractor on the Triple D were enclosed in a flexible rubber hose that passed between the two units. This photograph also shows the steering ram arrangement and the turntable and trunnion supports that were similar on both the Dual Power and Triple D tractors.

The Doe Triple D cost £1,950 in 1960 with a variety of wheel equipment available at extra cost. These special large-diameter steel wheels, supplied by J. J. Wright of Dereham, assisted traction but weakened the bull-gears and pinions and could cause rear-end failures.

The 12 x 38 tyres were usually specified as they gave the tractor extra clearance, better grip and greater speed. Darvill retractable strakes made by Stanhay were available for £50 a pair, while special narrow cage wheels to allow tight turning circles were offered at £84 for a set of four. The cage wheels were made in Germany by Kemper and supplied through a Colchester Tractors subsidiary, Colchester Tillage.

Special large diameter steel wheels to suit the Doe were supplied by J. J. Wright & Sons. Cranes of Dereham rolled the rims while Wrights welded Sankey centres into the wheels. Steels improved the machine's traction, but extended use weakened the bull-gears and pinions, and could result in expensive failures.

The Triple D was exhibited at the 1960 Royal Show held at Cambridge in July and came away with a silver medal.

The Triple D demonstration team. At the wheel, left to right, are Dick Partridge who built the ploughs, Ernest Charles Doe and demonstrator John Kent.

The Super Major version of the Triple D was launched at the 1960 Royal Show. It is seen working with a Doe toolcarrier and four-furrow plough with Doe's fitter, Dick Stevens, at the wheel.

From the end of that year, the tandem tractor was based on the new Super Major skid units and was able to take advantage of the important new features such as disc brakes, a differential lock and draft-control hydraulics that were introduced with the improved Fordson model. Another master cylinder and slave ram assembly allowed the differential lock on both the Triple D's power units to be synchronised.

Doe built most of the early Triple Ds out of two complete tractors, managing to sell off all the parts that were not needed through the trade. However, the surplus components became increasingly difficult to dispose of and by the time the Super Major model was in production, the company found it more cost effective to buy in industrial skid units for the front half of the machine. These were supplied from Ford minus front wheels, axles and hydraulics, and had a better radiator for improved cooling. The problem was that these radiators were more expensive

The Super Major Triple D was able to take advantage of all the new Fordson features, including disc brakes, differential lock and draft-control hydraulics. The tractor is seen working with a Doe folding toolbar.

This photograph, taken at the rear of the Ulting works, demonstrates the degree of articulation that could be achieved with a Triple D tractor. Dick Partridge is at the wheel.

A Super Major Triple D equipped with special narrow cage wheels that were designed to allow tight turning circles. The cage wheels were made in Germany by Kemper and supplied through Colchester Tillage, priced at £84 per set of four.

Based on two Super Major skid units, this Triple D tractor was once owned by the author. It is fitted with extra ploughing lights and a radiator guard. The leading tractor on many of the later Triple Ds was based on an industrial skid unit and had a more expensive radiator that was often protected by a fabricated guard.

is measured into a Triple D tractor that was undertaking a twelve-hour ploughing marathon on Lord Rayleigh's Farms at Terling Hall. The test took ›lace in August 1961 and the tractor used only 37 gallons of diesel to cover over 30 acres with a six-furrow plough during the twelve-hour period.

to replace, and after a number became damaged by drivers who had a habit of putting the tractor's nose into the hedgerow, the company fabricated a front guard that was supplied with several machines.

In August 1961, a Triple D tractor was subjected to a twelve-hour marathon test, working with a six-furrow plough on Lord Rayleigh's Farms at Terling Hall. Between 7 am and 7 pm, the tractor ploughed 30.9 acres and used 36.9 gallons of diesel. This averaged out at 2.57 acres per hour with a fuel consumption of 1.19 gallons per acre - figures that would put many a modern tractor to shame. The advantages of the Doe Triple D were plain for all to see, and sales rose steadily even though the price of the tractor had increased to £2,350.

The Doe Triple D was exhibited and demonstrated around the world. Tractors were exported across the globe to Israel, Nigeria, South Africa, Southern Rhodesia, Uruguay and Russia. In Scandinavia, Triple Ds went to

Dick Partridge leads five Triple Ds down the furrow during a demonstration of Doe tractors held in August 1961. The location is possibly Lord Rayleigh's Farms.

This Triple D tractor, seen outside the works at Ulting, was equipped with a Shawnee-Poole semi-trailer. The unit was ordered for road construction work on the M1 motorway extension.

Twenty-eight Triple Ds were exported, including this example on route to the Leipzig Show in Germany. The tractor is seen being transported to the docks by one of the company's Ford Thames Trader lorries.

Denmark and Sweden, while Austria, Germany, Spain and the Republic of Ireland accounted for sales in Europe. These export successes were a remarkable achievement for a modest family business and led to the company being asked to take part in the 1962 Lord Mayor's Show in London.

An interesting story relates to a Triple D that was sold to County Wexford in Ireland in September 1960. The customers were not prepared to pay the cost of having the tractor delivered by lorry to the port for shipment. To save money, they actually drove the tractor by road from Ulting in Essex to Fishguard in Wales. A journey of over 300 miles on a Triple D must have been an eventful experience!

A meeting in December 1961 at the Smithfield Show between Ernest Charles Doe and Ralph Christensen, a farmer from Badger in South Dakota, resulted in Triple Ds being shipped across the Atlantic for the first time. Christensen, who also worked for Owatonna Machinery, had already built his own Ford-based tandem tractor in 1958.

Christensen was evidently impressed with the Triple D, ordered one at the show and arranged for it to be shipped to the USA. This machine, tractor No. 210, was dispatched from Ulting on 8 December, arriving in America through the port of Houston in Texas in time to be exhibited at the National Western Stock Show in Denver. The Triple D

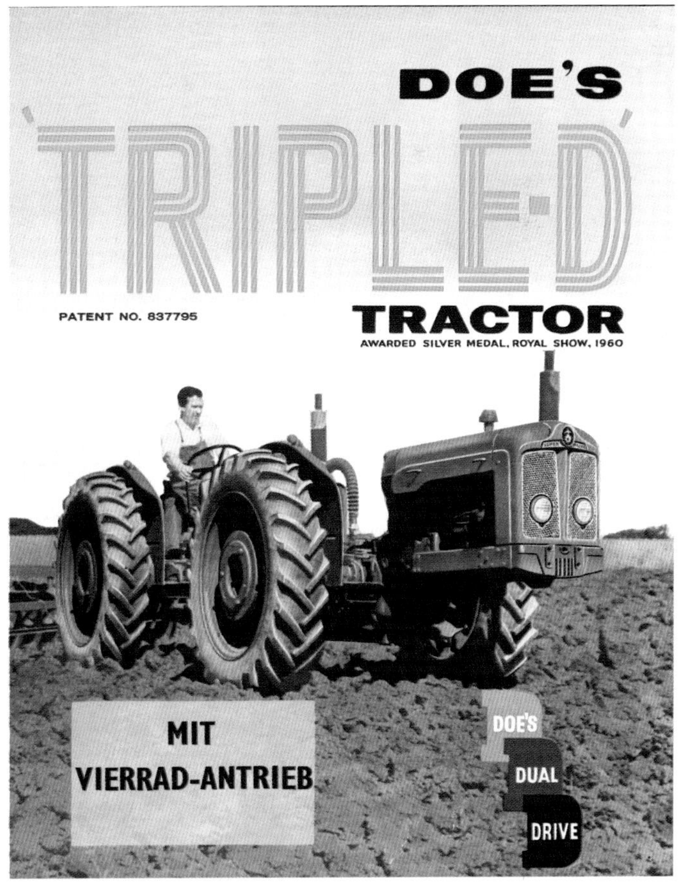

A German sales brochure for the Doe Triple D. The brochures were printed in several languages as the tractors were exported around the world.

A Doe Triple D is put through its paces on a Ford stand at the 1961 Herning Show in Denmark.

A brochure for the Double-T issued by Tandem Tractors, the American distributors of the Doe Triple D. The company, based in Selby, South Dakota, imported four tractors to the USA in early 1962.

created enough interest for Christensen to form a distribution company with two colleagues. The company was called Tandem Tractors Inc. and was based in Selby, South Dakota.

The tractor was marketed in the USA as the Double-T and was aimed at grain and rice growers. Three further machines were ordered from Ulting and were dispatched across the Atlantic during February 1962. These tractors, Nos. 216, 219 and 221, completed the consignment to the USA. No more were sent as Christensen came under pressure to concentrate on his work for Owatonna. The Double-Ts went as far afield as Colorado, Minnesota and New York State, and at least one still exists.

The Triple D was highly successful but it did have its drawbacks. Driving a machine nearly seven yards long that weighed over five tons was no mean feat on the road (as the author can personally testify). Problems with wear on the compensating ram, caused by many hours in one position when the tractor was working in a straight line, tended to lead to somewhat erratic steering when travelling at speed. Minute corrections of the steering wheel were often overcompensated for by the front end veering alarmingly across the white line or into the verge. It was not so much a case of steering the tractor as aiming it in the general direction of travel and hoping for the best! Blind junctions could be difficult, with the driver

This Doe Triple D was one of the four machines sold in the USA as the Double-T through Tandem Tractors. It is tractor No.216, one of the batch of three that were sent across the Atlantic in February 1962, and is seen in the hands of farmer Walter Bones working with an Oliver plough.

Driving a Triple D was no mean feat - the tractor itself was nearly 20 ft long without the added length of the eight-furrow semi-mounted plough. The photograph was taken in February 1964 on Orsett Estate, near Southend.

Dick Partridge demonstrates the manoeuvrability of a Triple D with a set of disc harrows. The tractor was surprisingly easy to handle with a little practice.

completely unsighted when trying to pull out into the road without hitting oncoming traffic.

In the field, the Triple D was very manoeuvrable, and surprisingly easy to handle with a little practice. The instruction book for the machine suggested: 'When using the tractor for the first time, it is advisable to drive around in a clear space until the driver is familiar with the controls'. An inexperienced operator, however, could easily get the two tractor units into two different gear ratios, or even have one in forward and the other in reverse gear at the same time. The results could be disastrous and expensive, but the tractor was fairly forgiving.

Derrick Hockley, at one time a service engineer for the Triple D, recalls being called out to a tractor that had reportedly 'completely seized up'. He was relieved to find that it was no more than a case of the driver having one unit in the high-ratio box while the other unit was in low gear. The wind-up between the two transmissions stalled both engines. No damage was done, but it was a salutary lesson for the operator to take more care.

Because the hydraulics and power steering were now both operated by the rear tractor, this unit tended to work harder. In fact, the two pumps drained around 5 to 6 hp. To compensate for this, the front engine was set to run 150 to 200 rpm faster so that it pulled harder and took some of the load off the rear unit. However, the rear engine always wore out more quickly than the front, particularly as most farmers, to save fuel, only used the back power unit when running on the road.

The Triple D was built for heavy cultivations, but the hard work sometimes took its toll on the turntable bearings and trunnion bushes.

The Triple D was very reliable provided it was regularly maintained in accordance with the company's recommendations. There were seven grease points on the turntable and trunnion that needed daily attention. Breakdowns were normally caused by abuse, misuse or poor maintenance.

Doe Triple D, Doe toolcarrier and Doe cultivator with Dick Stevens at the helm. The two large spanners, seen bolted together on the side-channel of the front tractor, were used to adjust the end clearance on the trunnion.

This Super Major Triple D has been fitted with the optional heavy-duty linkage to boost its lift capacity. Introduced in 1962, the linkage cost £100 and included two large assister rams. The tractor has a Fritzmeier cab and is seen with a Doe four-furrow reversible plough.

The company recommended that the clearance between the turntable faces was periodically checked and maintained by adjusting the centre-pivot nut. It has been suggested that failure to do this could result in the two tractor units parting company in the middle. Triple Ds did break in half, but this was usually due to abuse, misuse or poor maintenance, while bumping over fields at high speed did cause some frame fractures. It was also possible to buckle the rear wheel centres if the driver tried to turn at the headland before lifting the implement out of work.

The Triple D was supplied from new with two large spanners that were bolted to the left-hand side-channel on the front tractor. These were designed for adjusting the end clearance on the trunnion every three to four months. However, most users found that this operation was almost impossible outside the workshop because of the amount of clearance needed under the tractor to perform the job. It was also a two-man operation and probably best left to the service engineer.

Problems were encountered with movement of the

The last Triple D tractors to be based on the blue and orange Super Majors did not have Ernest Doe & Sons' trademark orange bonnets and were finished in normal Fordson livery.

hydraulic top-covers on the Major back-ends when using heavy implements, often resulting in damage to the internal oil-feed pipes and O-rings. Doe overcame this by fitting two extra dowels between the top covers and the main castings, and in 1962 offered an optional heavy-duty linkage assembly. It cost £100 and included two powerful booster lift-rams supported from an extended linkage.

A report by the National Institute of Agricultural Engineering raised doubts over the Triple D's suitability for working in hilly areas because the brakes only operated on the rear unit. However, this had little effect on sales as most of the tractors were bought to work on the intensive arable farms on the flat lands of eastern England. The original concept behind the Triple D made provisions for the owner to turn it back into two separate tractors for the summer months. In practice, few farmers bothered and Doe only ever sold one conversion kit of the necessary parts to for the changeover.

These few issues must not be taken out of context; they were relatively minor when compared to the performance of the tractor and the amount of work it did. Triple D breakdowns were actually more likely to be caused by problems with the Fordson skid units than faults with the Doe conversion itself - and it must be remembered that the Power and Super Majors were some of the most reliable tractors on the market!

The final manifestation of the Triple D, the blue/grey model based on the New Performance Super Major, was launched at the Royal Show at Stoneleigh in July 1963. The Super Major had a new Simms Minimec fuel pump and had been uprated to nearly 54 bhp. This power increase was reflected in the Triple D's new combined rating of 108 bhp.

Improvements heralded for the New Performance Triple D included heavy-duty bearings for the turntable and stronger trunnion bushes. There is some doubt as to whether these were actually altered, but stay-bars were fitted to strengthen the hydraulic lift mountings and reduce the load on the transmission housing. This was because the Super Major castings for the 'live-drive' model with the dual clutch were found to be not as strong as those used on the earlier models. A full hydraulic gear-change was included in the basic specification, along with 12 x 38 tyres and heavy-gauge wheel centres. Winsam or Fritzmeier cabs were available as optional equipment. The basic list price was now £2,450.

The first six Triple Ds finished to New Performance specification had a steel valance fitted beneath the front tractor unit to protect the sump and improve the machine's appearance. The valance was later dropped in the rush to

The New Performance Triple D was launched at the 1963 Royal Show. Based on the uprated 54 bhp Super Major, it was finished in the new Fordson blue/grey colour scheme and boasted a combined output of 108 bhp.

A sales brochure for the Doe Triple D based on the New Performance Super Major tractors.

get the tractors built as quickly as possible. Demand was at an all-time high due to the wet autumn of 1963. It became the busiest period of Triple D production with over sixty units leaving the Ulting works between July and December alone. The company was even planning to erect an 18,000 sq ft factory dedicated to tractor manufacture.

The performance of the Triple D became legendary. For a time, there was nothing else on the market to match it. It would cope with the worst conditions imaginable - and there is nothing much worse than Essex clay in a wet winter - and handle large implements that would tear lesser machines to bits. In the words of Derrick Hockley, who spent time acting as demonstrator for the Triple D, 'I have never driven a tractor that looked so clumsy yet did so much and worked so well.'

Further interest from behind the iron curtain led to Doe being asked to exhibit a Triple D tractor at the British Agricultural Exhibition in Moscow, opened on 18 May 1964 by the then Minister of Agriculture, Christopher Soames. Daily attendance figures as high as 10,000 were reported, with Soviet experts and technicians travelling from as far away as Siberia to view the machinery. Doe's stand received a visit from no less than President Khruschev himself, and the company secured the sale of its third tractor to Russia.

The eventual demise of the Triple D was brought about by the end of Fordson production at Dagenham. The loss of its skid unit meant that Doe had to re-think the design and come up with something different to suit the new Ford 5000 that was to be built at Basildon from late 1964.

Improvements brought in with the New Performance Triple D included the stay-bars that can just be seen below the number plate. These were fitted to strengthen the hydraulic lift mountings and reduce the load on the transmission housing. The twin top-links attached to the plough headstock.

Demonstrator Ben Golding puts the New Performance Triple D through its paces with a Doe four-furrow reversible plough. The steel valance seen under the hull of the front unit was only fitted to the first six tractors. It was discarded in the rush to get the Triple Ds built as quickly as possible when demand reached an all-time high during 1963.

A Highland pipe band heralds the opening of the British Agricultural Exhibition in Moscow. A Triple D was shown at the exhibition, and the top of one of its exhaust pipes can just be seen in the foreground. The banner displays the name of Ernest Doe & Sons Ltd. in both Russian and English.

The last Triple D, No.395, was dispatched to the Ford dealers D. T. Gratton & Sons Ltd. of Boston in Lincolnshire, on 3 October 1964. Sadly the production records for Doe's tandem tractors were partly destroyed and remain incomplete. Some confusion exists because the serial numbers do not match the actual numbers built, which is generally accepted to total 289 Dual Power and Triple D tractors dispatched between October 1958 and October 1964. No doubt when the figures were recorded, the company did not expect them to be still relevant or so carefully scrutinised over forty years later. Even today, the Doe family, although fiercely proud of their heritage, are still overwhelmed by the amount of interest shown in what is to them no more than a passing phase in their company's history.

President Krushchev (in the white hat, centre right) inpects a Triple D tractor, the bonnet of which can just be seen in the foreground, on Ernest Doe & Sons' stand at the British Agricultural Exhibition in Moscow. The show resulted in the sale of a third Triple D to Russia.

The Triple D's ability to cope with the toughest conditions made it a legend in its own lifetime. Nearly three hundred had been built when production ended in October 1964.

Chapter 4

DOE IMPLEMENTS

A Triple D with a six-furrow Doe mounted plough on Lord Rayleigh's Farms in 1961. On the left is the farms' tractor driver, F. Reeve, standing next to Doe's demonstrator, John Kent.

'The performance was outstanding and above all our best expectations .... our problem was to find suitable implements to use behind such power.' Alan Doe describes the launch of the Triple D and explains the dilemma the company faced after discovering that there were few ploughs or cultivators on the market to match the tractor's capability.

The only answer was for Ernest Doe & Sons to manufacture its own implements, and the first of these, a plough and a cultivator, were announced in 1959. The mounted plough was based on a Ransomes FR TS73 converted from four to five furrows. The implement was strengthened by a length of flat steel bar that was laid on its edge across the plough frame and clamped in place

The first Doe plough, introduced in 1959, was based on a Ransomes FR TS73 converted from four to five furrows. Note the strengthening bar clamped across the plough frame.

The Doe seven-tine heavy-duty cultivator was also announced in 1959 and was based on a box-section toolbar fitted with 24 in. depth wheels. Note that when the cultivator is at its working depth, the tractor's top-link is horizontal and vulnerable to breakage should the implement encounter an obstruction.

The Doe seven-tine cultivator was designed to work to a depth of 18 in. and was ideal for pan-busting.

with U-bolts. It was available with Ransomes YL, EPIC or FRDCP bodies and cost £225, complete with assister rams to boost the tractor's lift capacity.

The seven-tine heavy-duty cultivator was priced at £230 and was based on a box-section toolbar fitted with 24 in. depth wheels. It was a very strong piece of equipment, designed for cultivating up to 18 in. deep and was ideal for pan busting. It could be adapted to take nine tines, two 22 in. subsoiler blades or a mole drainer attachment.

Some problems arose because when the cultivator was working at its full depth, the tractor's top-link was horizontal and vulnerable to breakage should the implement encounter an obstruction. After several bent top-links, Doe began manufacturing its own heavy-duty link. This proved costly to make because of the high-

The Doe toolbar could be adapted to take two 22 in. subsoiler blades. John Kent is at the wheel of a Power Major Triple D.

A Doe seven-tine cultivator pictured at the rear of the Ulting works. This is a later version with a revised headstock, designed to alter the geometry of the tractor's linkage to prevent damage to the top-link.

tensile steel and the expense involved in turning both left- and right-hand threads. It was priced in the region of £40 to £50, but a good number were sold. The cultivator's headstock was eventually modified to provide extra link positions and alter the geometry of the linkage so that the lower link-arms ran parallel to the ground with the top-link at an angle.

The following year saw the launch of the Doe telescopic toolbar consisting of a heavy-duty box-section bar with two extendable sliding ends, each fitted with its own three-point linkage attachments. This arrangement allowed Ransomes Duchess disc harrows or Danish Flemstofte spring-tine cultivators, the latter supplied through Colchester Tillage, to be hitched in pairs behind the Triple D to give working widths of up to 20 ft.

The first Doe reversible plough was unveiled in prototype form at the Royal Show in July 1960 with a price tag of £550. It was a four-furrow mounted model with a tubular frame and hydraulic turnover using a rack and pinion mechanism. A revised production version of the plough was demonstrated later the same month at Boyton Hall Farm. It had a longer tubular main-beam, now cranked to give a greater clearance between the bodies, and an improved turnover mechanism. A new heavy-duty four-furrow version of the conventional Doe plough, fitted with Ransomes SCN bodies and priced at £215, was seen at the same demonstration, along with a heavy set of trailed disc harrows.

Launched in 1960, the Doe telescopic toolbar is seen fitted with a pair of Danish Flemstofte cultivators. The toolbar had two extendable sliding ends, each with its own set of three-point linkage, to allow working widths of up to 20 ft.

The first Doe reversible plough was introduced at the 1960 Royal Show, priced at £550. It was a four-furrow mounted model with a tubular frame and hydraulic turnover.

The early tubular-beam reversible plough is seen behind a 1961 Super Major Triple D on Lord Rayleigh's Farms. This first Doe reversible was not a great success. It proved difficult to set and only twenty were built.

The six-furrow version of the conventional Doe plough was introduced in 1961. It is seen on a Super Major Triple D working on Lord Rayleigh's Farms.

The first reversible plough was not a great success and proved difficult to set due to distortion of the beam caused by the welding process during manufacture. Only twenty were built before it was redesigned during 1962. The result was an improved four-furrow reversible with a bar-frame and utilising several Ransomes parts. It was available with SCN, YL, EPIC or FRDCP bodies. A six-furrow version of the conventional plough had also been introduced during 1961. Both of the new ploughs required the Triple D to be fitted with the heavy-duty linkage assembly incorporating two booster lift-rams. A double-acting ram (DAR) valve assembly was also needed to operate the reversible plough's hydraulic turnover mechanism.

The four-furrow reversible plough was available with or without a depth wheel according to customers' requirements. The depth wheel could be positioned to run on the land, or at the rear to run in the furrow. The rear depth wheel was usually used in conjunction with a toolcarrier, but if the plough was mounted directly to the tractor, then a telescopic top-link was needed.

The implement range was extended still further through 1963. The Doe Show in February saw the launch of a fifteen-tine ripper plough - a universal cultivator

An improved Doe four-furrow reversible plough was introduced during 1962. It had a conventional bar-frame and utilised several Ransomes parts. Note the rack and pinion turnover mechanism.

The Doe four-furrow reversible plough was fitted with Ransomes bodies and was designed to be used with the heavy-duty linkage assembly as seen on this New Performance Triple D.

A four-furrow reversible plough with the rear-mounted depth wheel. The rear depth wheel was usually used in conjunction with the toolcarrier, but if the plough was mounted directly to the tractor, then a telescopic top-link was needed. Note the rods running alongside the frame to pull the depth wheel round as the plough was turned over. The tractor is Ernest Doe & Sons' own restored 130 model.

Many of the later Doe reversible ploughs were finished in a metallic green-grey colour. Most of the conventional mounted ploughs were blue while the steerable semi-mounted ploughs were painted orange.

A Super Major Triple D at a demonstration with a Doe fifteen-tine ripper plough. Introduced in 1963, this universal cultivator was designed for medium-duty use such as stubble busting. A Power Major Triple D can just be seen on the right of the photograph.

Ernest Doe & Sons introduced this five-tine cultivator at the 1963 Essex Show. The implement was designed for 50 hp tractors and is seen behind a Fordson Super Major.

with a box-section frame. Priced at £245, it was designed for lighter use than the seven-tine cultivator and was ideal for stubble busting or aerating soil. A new telescopic toolbar/Flemstofte spring-tine combination with an increased working width of 22 ft was introduced at the same time.

The company had a massive stand at the Essex Show in June, and new additions to the implement range on display included a five-tine ripper for smaller 50 hp tractors, a three-furrow plough and a hydraulic folding toolbar for the Triple D. The three-furrow was a large clearance plough fitted with SCN bodies. It would turn in 16 in. furrows and was suitable for burying straw or trash. Several were sold to the green and salad growers in south Lincolnshire and the Fens, where they were used to plough in the residue of crops such as celery.

The folding toolbar was designed to carry two disc harrows or two spring-tine cultivators. The tractor's hydraulics operated a pair of rams that automatically folded the toolbar from its working width to a travelling position, the whole operation being controlled from the driver's seat. This

The Doe three-furrow reversible was a large clearance plough that was fitted with SCN bodies and would turn in 16 in. furrows. Also launched at the 1963 Essex Show, it was ideal for burying straw or trash.

enabled the implement to pass through gateways or travel safely on narrow roads. Many of the folding toolbars were made by an outside contractor at Bishop's Stortford. A few of these gave problems as the ram mechanism sometimes went over-centre and would not unfold.

The largest plough that Doe ever built, an eight-furrow model, appeared at the 1964 Doe Show. It was a semi-mounted steerable plough with a swivelling headstock. A hydraulically-controlled depth wheel allowed the plough to be progressively lowered into work to leave a tidy headland. The front body of the plough was lowered into the ground first, followed one

The Doe folding toolbar had two sets of linkage and could carry two disc harrows or two spring-tine cultivators. A pair of hydraulic rams folded the toolbar from its working width to a travelling position. It is seen with two 9 ft sets of Ransomes disc harrows in the folded position.

The Doe folding toolbar equipped with a harrow frame. The implement is seen behind a Roadless 6/4 Ploughmaster tractor.

The largest plough that Doe built was an eight-furrow model introduced in 1964. It was a steerable semi-mounted plough with a hydraulically-controlled depth wheel. This was the original version that was used without a top-link.

Later versions of the Doe steerable semi-mounted plough had a revised headstock with two top-link mountings. This is a six-furrow version of the 7/6/5 model. Note the single link-rod connected to the front of the headstock. This bolted to the tractor's right-hand lower link-arm for adjusting furrow width.

at a time by the other bodies as the ram on the depth wheel slowly closed, all controlled through a restricter in the tractor's DAR valve.

The original steerable plough did not use a top-link, but this arrangement was found to be unsatisfactory and a rectangular headstock with two top-link mountings was soon introduced. When this was used in conjunction with the heavy-duty linkage, two top-link rods were fitted. A further single link-rod could be bolted to the right-hand lower link-arm for adjusting furrow width.

The semi-mounted plough was designated the 8/7/6 and could be converted to seven or six furrows by the removal of one or two leg assemblies. To convert it into seven furrows, the front leg assembly was removed and

The 5/4 version of the steerable plough. The plough has been converted into a four-furrow by the removal of the rear body. Note that the depth wheel and ram assembly has been relocated in a forward bracket further along the beam.

A Doe 7/6/5 steerable semi-mounted plough seen loaded aboard a lorry at Ulting in September 1965. The plough was part of a shipment destined for East Germany.

the plough offset by swivelling the headstock through 180 degrees. To reduce the plough to six furrows, the rear body was removed and the depth wheel and ram assembly was relocated in a forward bracket. Other models in the same range included the 7/6/5 and 5/4 furrow versions, all using YL, EPIC or SCN bodies.

The implements were all manufactured in Doe's workshops with most of the work carried out by fitters Dick Partridge and Ernie Allen. Any fabricated or welded parts, including the main frames, were sent to annealing

The tool carrier, seen at the rear of the Ulting works, was introduced in 1964 to allow crawlers and other wheeled tractors to use Doe implements. It was basically a heavy-duty three-point linkage arrangement that was carried on a rectangular main frame mounted on wheels. It is shown equipped with a Doe folding toolbar.

Dick Stevens is seen at the controls of a Track-Marshall 70 with a Doe tool carrier and seven-tine cultivator. The tool carrier cost £300 with a hydraulic kit to suit Track-Marshall crawlers available for an extra £180.

The tool carrier enabled crawler tractors to handle heavy equipment such as this four-furrow Doe reversible plough. There was a certain amount of co-operation between Ernest Doe & Sons and Marshalls of Gainborough, and this 70 crawler was Track-Marshall's own demonstration tractor.

A Track-Marshall 70 on demonstration with a Doe tool carrier and a three-furrow reversible plough in 1966, possibly at the Doe Show. It was a combination that was particularly popular in south Lincolnshire and the Fens.

furnaces at March in Cambridgeshire to be 'stressed' by heat treatment before being returned to Ulting for painting. The expanding line of Doe farm implements began to attract attention from farmers who needed heavy-duty equipment but were not necessarily owners of Triple D tractors. To tap into this market, the company introduced a tool carrier to enable crawlers and other wheeled tractors to be able to use Doe reversible ploughs and heavy implements.

The tool carrier consisted of a rectangular main frame mounted on wheels equipped with pneumatic tyres. It had a heavy-duty three-point linkage system, operated by two single-acting rams together with a slewing drawbar that allowed the tractor to plough on-the-land.

Dick Stevens at the wheel of a Super Major Triple D working with a Doe tool carrier and a four-furrow reversible plough. The tool carrier had a slewing drawbar that allowed the tractor to plough on-the-land.

The carrier was launched at the 1964 Essex Show and cost £300 complete with all the necessary hydraulic hoses. For tractors without hydraulics, a kit consisting of a Vickers V-20 pump, a tank and filter, hoses and control valves was available for an extra £180. The hydraulic kit was designed for Track-Marshall crawlers, but could be adapted to suit other makes.

The launch of the replacement for the Triple D, the Doe 130, in December 1964, saw little change to the implement line except for small modifications to accommodate the new tractor's extra horsepower. The working width of the telescopic toolbar/Flemstofte spring-tine

A Triple D in action with a tool carrier and a Doe seven-tine cultivator. One advantage of the tool carrier was that it took the weight of the heavy implement off the tractor and did away with the need for the heavy-duty linkage.

The launch of the more powerful Doe 130 in late1964 saw little change to the implement line. A Doe four-furrow reversible plough is seen behind a 130 tractor at the sugar beet autumn demonstration, held near Hoxne in Suffolk in October 1965.

The steerable plough range was revised during 1966 when Ernest Doe & Sons introduced its own plough bodies, manufactured for the company by Lemken. The type NM/6/12 plough is seen behind a Doe 130. The additional steel depth wheel was designed to keep the plough on an even plane in work.

combination was increased to 31 ft, and a new 25 ft Bjurenwall spring-tine cultivator that folded for transport was imported from Sweden through Colchester Tillage.

Problems in sourcing Ransomes parts, with supplies sometimes taking up to four months, led to Doe introducing its own plough bodies at the 1966 Royal Show. The bodies were manufactured for the company by Lemken, the obvious choice of supplier as its products were marketed in the UK through Ernest Doe's associated company, Colchester Tillage. The DHS (Doe High Speed) 1 was a general-purpose body similar to a Ransomes YL. The DHS2 was a cross between a general purpose and a semi-digger body while the DHS3 was a full semi-digger. The DHS2 was very successful and was eventually adopted by Lemken as the basis for its LWS body.

The steerable plough range also came in for some changes with a revised headstock and the option of hydraulic furrow-width adjustment. Three models were available; the type S5/14 was a 14 in. five-furrow plough. The other two models were designed to turn 12 in. furrows;

A Doe 130 working with a fifteen-tine ripper plough in Lincolnshire during 1967. This combination could cultivate well over three acres of land per hour.

the type NM/6/12 was a six-furrow, convertible to five or seven, while the L8/12 was an eight-furrow that could also be converted to a seven-furrow.

Doe demonstrated a new Universal steerable semi-mounted plough at its own show in February 1967. The Model U was designed as a basic 'high-speed' plough that could be reduced from five to four or three furrows by removing one or two rear legs and repositioning the hydraulic depth wheel assembly further along the box-section main-beam.

The Universal, or Model U, steerable semi-mounted plough was introduced at the 1967 Doe Show. The body assemblies were clamped to the beam and could be removed or repositioned to allow different furrow widths.

Different furrow widths could also be obtained by sliding the body assemblies along the beam. This arrangement was not entirely successful as the clamps holding the body assemblies were found to work loose and most ended up being welded into place.

The same plough could be converted for on-the-land ploughing by using a special headstock in conjunction with an extension to the front cross-beam. The advantage of on-the-land ploughing was that it caused less soil damage and put less stress on the linkage. Prices for the plough started at £321 10s. Other ranges of Doe equipment launched the same year included new models of subsoilers, mole drainers, ripper ploughs, spring-tines and the M-type chisel plough for heavy cultivation.

Roadless Traction found the Doe ploughs an ideal match for its own high-horsepower four-wheel drive tractors and even offered its own special headstock to suit for £24. Introduced during 1967, the modification allowed sufficient offsetting for the tractor to plough on-the-land without any side-draught effect.

Ernest Doe & Sons also branched out into accessories and attachments for Ford tractors, including auxiliary lift rams, pick-up hitch drawbars and rear-wheel scrapers. The lift-ram kit for Ford tractors, with a single acting-ram, hose and fittings, cost just under £40. It was designed by Derrick Hockley,

A Doe Universal plough at work behind a Track-Marshall 70 crawler.

for which he received a £10 bonus from the company. Doe even marketed a cab heater that used hot water from the tractor's cooling system circulated through an electric fan assembly to blow warm air around the driver's legs.

By the mid-1960s, the retail side of Ernest Doe & Sons' business was booming. The company's manufacturing activities, however, were not so profitable and were taking up valuable workshop space. The decision was made to wind down tractor and implement production, and few machines were made after 1968.

The Universal plough could be converted for on-the-land ploughing by using a special headstock in conjunction with an extension to the front beam.

A Roadless Ploughmaster 95 working with a Doe plough. Roadless Traction introduced its own headstock to suit Doe ploughs in 1967. It cost £24 and allowed the tractor to plough on-the-land without any side-draught effect.

Chapter 5

DOE TRACTORS

An early Doe 130, equipped with a fifteen-tine ripper plough, photographed near the Ulting works in 1965.

Sunday 10 October 1964 was an important milestone in the history of Ford tractor production. It was the day that the company unveiled its new 'worldwide' range of tractors at Radio City Music Hall in New York. Attending the launch were nearly 6,000 dealers representing the tractor markets of 120 different nations of the world. Ernest Charles Doe was among that delegation.

It was a lavish launch conducted with a musical revue, orchestra and dancing girls from the famous Rockettes troupe. But the pizzazz only served to underline the importance of the new integrated range of tractors, designed for simultaneous worldwide assembly at plants in Basildon in England, Antwerp in Belgium and Highland Park in the USA. The new range, codenamed 6X, would replace both the existing American tractor line and the Fordson Super Major and Dexta models at Dagenham.

The flagship model in the 6X range was the Ford 5000, a 65 bhp tractor with an eight-speed gearbox. It was the direct replacement for the Super Major and was the machine that was of most interest to the specialist conversion builders, such as County, Roadless and, of course, Doe. The 5000 represented a great leap forward in design and incorporated several features that improved on the Super Major, but it was a completely new tractor requiring Doe and their fellow specialist equipment manufacturers to redesign or adapt their machines to take the new skid unit. The work had to be carried out in a very short time if these companies were to be ready for the UK public launch of the 6X range at the Smithfield Show in December.

The brochure heralding the arrival of the Doe 130. The tractor was introduced at the 1964 Smithfield Show with a list price of £2,850.

Doe's tandem tractor was a relatively simple design and it was not too difficult to adapt it to suit the Ford 5000 skid. New sub-frames were made, and for the first time the company introduced jigs to ensure accuracy of manufacture - an improvement on earlier methods as it has often been asserted that no two Triple D frames were ever the same!

The trunnion assembly was strengthened and a new cast-steel turntable incorporating two rows of steel ball-bearings was fitted. Because the new eight-speed gearbox had two gates, an electric

The Doe 130 was based on two 65 bhp Ford 5000 skid units making it a 130 bhp tractor. It had a new sub-frame, a strengthened trunnion assembly and a new cast-steel turntable. Dick Stevens is in the seat.

solenoid was added to the hydraulic servos on the front slave unit to move the gear lever across to the fourth/eighth gear position. The steering was changed to hydrostatic with an orbital valve that did away with the need for the compensating ram and linkage. Ernest Doe also took the opportunity to simplify the design of the steering arrangement and reduce manufacturing costs by using only two double-acting slewing rams instead of four. Finally, the tractor units were fitted with heavy-duty 13 in. clutches in place of the standard 12 in. plates to cope with the increased stresses.

The steering arrangement fitted to the early Doe 130s consisted of just two rams - one either side of the turntable. One ram pushed while the other ram pulled. It was not a successful arrangement because it put undue strain on the power steering pump, seen mounted on the front of the rear tractor unit.

The new tandem tractor, the replacement for the Triple D, was launched as the Doe 130 at the Smithfield Show in December 1964. The tractor's numerical designation reflected the combined 130 bhp of the two Ford 5000 units. The 130 was priced at £2,850 and drew great interest from prospective customers already aware of the proven pedigree of the tandem tractor design and attracted by the extra 30 hp offered by the new machine.

Seventy-three Doe 130s were sold during 1965, fourteen of which were exported. A batch of eight machines was ordered by Messrs Campbell Booker Carter for the Nigerian Sugar Cane Company. They were to be used for hauling sugar cane trailers from the field to the refinery and represented the largest single order for Doe tractors.

The Nigerian 130s had different sub-frames to give them greater ground clearance when working in the harvested crop. Instead of having the usual frames going underneath to the bell-housing, the front units were supported by gusset plates mounted over the top of the transmission. The tractors were also fitted with special front guards to protect the radiators and had their gear selectors altered to blank off top-gear so that any possible damage through speeding over rough ground could be avoided.

Another Doe 130 was ordered by a farmer in Australia. The sale was handled by the Queensland dealers, Toft Brothers of Bundaberg, who delivered it to the farmer, A. J. Coaker of Narromine, New South Wales. The tractor was supplied with extra-wide fenders and had heavy-duty wheel centres with 15 x 30 tyres.

Because the Australian 130 was going to be used with trailed implements on an intensive wheat farm, no hydraulic lift or linkage was fitted, but the tractor was equipped with

This batch of eight 130 machines destined for Nigeria represented the single largest order for Doe tractors. Exported during 1965, they were used for sugar cane work and had top-gear blanked off. Note the special guards to protect the radiators. Len Cordy, Ernest Doe & Sons sales manager, is on the first tractor on the left. Fitter George Delgado is at the wheel of the first tractor in the right-hand line-up with office manager, Michael Taylor, behind him.

The Doe 130 was very popular with large-scale arable farmers. It was one of the most powerful tractors on the market and there was little else to touch it in terms of performance and reliability.

only the one Select-0-Speed 130 was built and was sold locally to Eastlands Farm at Bradwell-on-Sea.

The 1966 Royal Show also saw the introduction of a new heavy-duty lift assembly for the 130 tractor. The assembly consisted of heavy-duty linkage, two hydraulic assister rams, a pump and a control valve unit. The auxiliary hydraulic pump was mounted on the side of the rear engine and driven off the camshaft. The control valve allowed the pump to kick in automatically once pressure built up in the tractor's own hydraulic lift system. The complete assembly cost £160.

This Doe 130, fitted with 15 x 30 tyres and a Fritzmeier cab, was photographed when still at work in 1983.

Exactly 170 Doe 130s were built and the very last to leave Ulting was delivered on 29 May 1968. By happy coincidence, the customer was George Pryor.

The days of the tandem tractor were becoming numbered as the company faced competition from home and abroad. The County 1124 and the Hungarian Dutra D4KB were already established on the market, offering four-wheel drive and a similar horsepower and tractive ability to the Doe for less money. Other high-horsepower four-wheel drive contenders, including the Muir-Hill 101, the Northrop 5006 and the Roadless 115, were equally attractively priced. Unlike the Doe, they were all compact machines with only one engine to fuel and maintain. The tandem tractors had filled a niche in the market when there was nothing else available, but now they were becoming outdated if not outclassed.

In April 1968, Ford introduced its revamped 6Y tractor range, dubbed Ford Force, with major improvements to the 5000 model in answer to criticism of its lack of power and torque. The new 6Y 5000 boasted 75 bhp and new cleaner styling. Doe incorporated the new skids into its tandem tractor and re-launched it as the Triple D 150.

The 150 was advertised as a 'New Style Tractor - Better, Quicker, Cheaper', which rather

This nicely restored 1967 Doe 130 tractor belongs to Burtts of Dowsby.
This Lincolnshire farming family once operated a Triple D during the early 1960s.

An aerial photograph of Ernest Doe & Sons yard and works at Ulting, taken in the mid-1960s. The large building in the centre right of the photograph is the tractor assembly shop with skid units ready to be made into Doe 130s lined up outside. The 130 demonstration tractor can be seen in the foreground, just behind a County with a Doe plough. A Northrop is in the same compound, parked at the end of the line of Ford tractors near the road. Note all the JCB equipment on site.

The Doe 150 that replaced the 130 model in 1968 was based on two 75-hp Ford 'Force' 5000 tractors. It was only in production from July to September and only three were built. This example owned by Steve Haylock from near Cambridge has had its power units replaced but is believed to have been built from an original 150 frame.

contradicted its price tag of £4,200. It was actually no different from the 130 apart from the power increase and styling changes to the Ford units with new grille inserts, different exhaust pipes and full-length decals. It sadly proved to be the last Doe tandem tractor and only three were built.

The first Doe 150 was sold in July 1968 to a farm in Lincolnshire. The second machine went to a ploughing contractor from Wilburton near Cambridge in August, while the third and last was delivered to Lord Rayleigh's Farms in September.

The demise of Doe's tandem tractor designs was not just due to falling demand. Proposed new safety cab laws, due to take effect in September 1970, were causing the company some headaches. Ford's safety cab was tested for its own tractors, but it was doubtful whether it would be automatically approved for the combined weight of the tandem machines. Rather than become involved in costly legal wrangling or go through expensive cab approval processes, Doe thought it simpler to drop its tractor from the market and concentrate on the more profitable retail side of the business.

Lack of profit was probably the overriding factor in the decision to cease tractor production. Doe's pricing structures were designed to be as competitive as possible and this left little return for the considerable investment involved. Alan Doe openly admits that his company made little profit from the Triple D and its successors. He is quoted as once saying, 'I am sure that we are better at distributing machinery rather than manufacturing it.'

Plans did exist for a number of improvements to the Doe tandem tractor had production continued. A 130 that is still owned by the company was used as a test-bed and had a strengthened turntable with three rows of ball-bearings. Different gear-change mechanisms were being considered, including an all-electric system or an air-over-hydraulic arrangement with a compressor fitted to one of the engines. Neither was developed very far.

A rare sighting of a 188 hp Doe tractor based on two Ford 7000 skid units. The machine started life as a 130, but was one of several that were later rebuilt at Ulting at the request of customers. It is seen in Frank Foot's yard near Winchester in 1979, prior to being broken up for export.

Ernest Doe & Sons' own Doe 130, seen in action at the 2001 Doe Show, was used as a development tractor. It was a test-bed for proposed improvements that would have been introduced had Doe tractor production continued and has a special turntable with three rows of ball-bearings.

Alan Doe believes that had the company continued with its tandem tractor designs, then the logical step would have been to keep the Ford 5000 unit at the front and use a turbocharged Ford 7000 skid as the back tractor, the idea being to compensate for the power loss experienced by the rear engine. Thc combination would have been called the Doe 170. As it was, the 150 model delivered in September 1968 represented the last new Doe tandem tractor to leave Ulting.

Doe tractors continued to work for many years to come. They were invaluable for heavy cultivations and were almost unbeatable when the conditions turned bad. Sightings of Doe 130s keeping pea viners moving around the clock after periods of heavy rain remained a regular occurrence for many years in the eastern counties.

Committed Doe users believed that there was nothing to touch the tandem tractors in terms of performance, and machines were regularly returned to the workshops at Ulting to be reconditioned or rebuilt. Several were uprated to 150 specification, with the early 6X tractor units on the 130s replaced by the more powerful 6Y skids, while one or two were made into 5000/7000 combinations. At least one 130 tractor was rebuilt with two 7000 skid units, but it is not known whether this was done at Ulting or in a farm workshop.

The end of Doe's tandem tractor production coincided

Even after Doe ceased production of its tandem tractors, several 130s were returned to Ulting to be reconditioned or uprated to later specification. This combination of Ford Force 5000 (front) and Ford 7000 (rear) power units was a successful arrangement. The registration plate indicates that the tractor was a 130 originally built in 1967.

The late-1960s saw an upsurge in Ernest Doe & Sons' retail business and Ernest Charles Doe concluded a deal worth £90,000 for ninety new Ford Force tractors from the Ford Motor Company on the opening day of the 1968 Royal Show. Some of these tractors are seen arriving at Ulting from Basildon. Office manager Michael Taylor and salesman Derrick Hockley check driver Joe Oliver's consignment note.

with the end of its implement manufacturing activities, but this was more than compensated for by the upsurge in the company's retail business and its success in selling the Ford tractor range. On the opening day of the 1968 Royal Show, Ernest Charles Doe placed an order for ninety of the new Ford Force tractors. This was the single largest order ever received by Ford Tractor Operations and was worth over £90,000 at the time.

Doe had increased its sales territory for Ford tractors, being awarded a main dealership franchise for Cambridgeshire in October 1966. The company operated from premises at Fulbourn, about five miles from Cambridge. The one-acre site was a former mill that was converted into showrooms, workshops and a parts store. A further Cambridgeshire branch at Littleport was added in 1970. These two new outlets joined existing Doe branches at Fyfield, Rochford, Stansted, Braintree, Sudbury and Dartford. The Dartford branch in Kent had been opened in 1965 after Alan Doe successfully negotiated the purchase of the site, the Darenth Valley Iron Works in Sutton-at-Hone, from the old-established horse

Ford Force tractors, a New Holland baler and the branch's Ford 400E service van on the forecourt of Ernest Doe & Sons' Fyfield depot in 1968. The company opened new branches in Cambridgeshire and Kent between 1966 and 1970.

The Doe 5100 was introduced in 1970 to meet the demand for a high-speed two-wheel drive with a six-cylinder engine for pto work and high-speed cultivations. It was based on a Ford 5000 fitted with a 98 hp Ford 2704E industrial engine supported by a cast-iron hull supplied by EVA of Belgium.

John Kent demonstrates the Doe 5100 in October 1970. The tractor was assembled at Ulting and was fitted with power steering, an assister ram and 14 x 34 tyres as standard. Its list price was £2,850.

The Doe 5100 in action with a three-furrow Ransomes reversible plough. Only five or six were built because the introduction of the tractor was met with an adverse reaction from the Ford Motor Company. There was also some concern over how the crown-wheel and pinion might stand up to the power of the six-cylinder engine.

hoe and cultivator manufacturers, G. J. Garrett & Sons. Gosfield was closed in 1967 and its business transferred to the larger Braintree branch nearby.

The company was never one to miss an opportunity, and was reluctant to completely break away from its tractor building activities when it identified a gap in the market that it felt it could plug. Feedback from several of its customers convinced Doe that a more powerful two-wheel drive tractor was needed at the top end of the Ford range. The 6Y 5000 was a good machine, but its 75 hp was not enough for some operations such as cultivation and silage work.

What was needed was a tractor with greater output at the power take-off to handle the latest power harrows and high-capacity forager harvesters. The obvious solution was to slot a six-cylinder engine into the 5000 chassis. County and Roadless had been carrying out similar transplants for years for their four-wheel drive machines, but both were

The Doe 5100 is a rare tractor, but at least two exist including this example still working in Lincolnshire. It is seen fitted with Terra-tyres on Manby airfield in 1991.

reluctant to upset their supply agreement with the Ford Motor Company by offering competing two-wheel drive models in the UK. Doe evidently had no such concerns and began to source components for its own conversion.

Engines were no problem as the 2703E and 2704E industrial six-cylinders were built at Dagenham and were readily available through Ford's Industrial Power Products division. Doe plumped for the 5,899 cc (360 cu in.) 2704E power unit rated at 98 hp. The drawback was that the engines had unstressed blocks and aluminium sumps and needed to be supported by a sub-frame or side-channels for use in a tractor.

Alan Doe's son, Colin, is the current managing director of Ernest Doe & Sons, appointed to the position in 1989. He is seen here with a line of New Holland tractors in the yard at Ulting.

Doe got round this by using a cast-iron hull supplied by EVA of Belgium. Based in Andenne, EVA was run by Willie Everard, a Ford dealer and importer who had built up a relationship with Ernest Doe & Sons through buying second-hand tractors from the Essex company. Everard had developed the hull in 1965 to act as a sub-frame for his own six-cylinder conversions of 5000 tractors that were supplied through Ford dealers in Belgium, France, Germany and Holland.

Doe's six-cylinder tractor was assembled at Ulting. A special flywheel and clutch assembly was supplied by County, and the other parts, including a longer bonnet and extended drag-link, were made or modified in the workshop. The battery had to be repositioned from under the bonnet and was mounted on the right of the bell-housing in front of the footplate. Doe's own hydraulic assister-ram kit was fitted, and power steering and 14 x 34 tyres were standard.

The first six-cylinder tractor was demonstrated in September 1970. It was a less-cab model and was badged as the Doe 5100. It was officially launched at the Doe Show in February 1971, priced at £2,850 to include a Duncan safety cab to meet the latest regulations.

The 5100 was advertised as a 100 hp tractor 'ideal for high-speed cultivations and pulling heavy trailers in addition to many other jobs on the farm'. Considerable interest was shown by West Country farmers who were attracted by the tractor's pto power for forage work. Unfortunately, however, the launch of the 5100 met an

Ernest Doe & Sons' main franchise became New Holland following the 1991 merger of the Ford and Fiat agricultural and construction machinery divisions. One of Doe's demonstration tractors, a New Holland TM165, is seen at the Doe Show in February 2000.

Ernest Doe & Sons' yard at Ulting in March 1998. Compare this with the photograph on page 83; the road has been realigned and the main works building has been replaced with a new structure. Most of the older buildings still survive, including the ex-bus depot hanger and Ernest Doe's house, Hill View, now part of the office complex. The original blacksmith's shop would have stood in the area covered by the buildings in the bottom left of the photograph.

adverse reaction from the Ford Motor Company.

Ford Tractor Operations had always been opposed to six-cylinder conversions as it was felt that they could shorten the life of the transmissions. More importantly, Ford had its own 100 hp machine in the pipeline; the new 7000 with a 94 bhp turbocharged four-cylinder engine was due to be released at the 1971 Smithfield Show and the company did not want to face competition from one of its own dealers.

Gentle pressure was put on Ulting to cease production, and rather than risk its good relationship with Ford, Doe quietly dropped the 5100 from the listings. It was not a difficult decision to make; Doe was already experiencing problems in persuading Ford to supply it with skid units minus engines, and was also having to offer its own warranty for the completed tractors. There was some concern over whether the 5000's crown wheel and pinion would stand up to the power of the six-cylinder, and it was beginning to look as if the project could become an expensive exercise for the company. Only five or six 5100 tractors were built. At least two still exist, but the whereabouts of any others are unknown.

The 1970s and 80s was a period of consolidation for Ernest Doe & Sons as the company built on its success to become one of the largest machinery retail organisations in the UK. Doe's franchise for Ford industrial equipment extended across the whole of south-east England, and branches at Bermondsey, Mitcham and Send in Surrey, and Watford in Hertfordshire came and went. Another construction equipment depot at Ashwellthorpe in Norfolk was eventually replaced by a larger branch at Wymondham. It must not be forgotten that the Doe family also controlled Colchester Tractors, Colchester Tillage, Doe Motors and Doe Finance.

Alan Doe succeeded his father as chairman in 1971, and Ernest Charles became life president until his death in 1979. Alan Doe's son, Colin, was appointed managing director of Ernest Doe & Sons in 1989 and began laying the foundations to take the company through into the twenty-first century. The grass machinery and professional turf care business was increased, a tractor hire fleet was established and the farm machinery business expanded into Norfolk by offering agricultural sales and service from the Wymondham site. The company also acquired the former Bracey's of Benington depot in Hertfordshire from Dalgety in 1990.

The construction equipment division was strengthened by the addition of Komatsu, Furukawa, Winget, Barford and Hyundai franchises. The company also marketed the Doe 12000 MDT articulated dump truck. This 12 tonne four-wheel drive machine was designed and manufactured by MT Engineering of Braintree and based on a Ford 7710 skid unit.

Doe's main franchise underwent a number of changes in 1991 following the creation of New Holland from a merger of the Ford and Fiat agricultural and industrial machinery divisions. Ernest Doe & Sons became New Holland dealers and was able to take advantage of the new organisation's extensive global product line.

These machinery displays formed part of the 40th Doe Show, held in February 2000. The show regularly attracts visitors from all over the UK and has been held every year since 1960.

Colchester Tractors was brought back into the fold in 1993. The business was purchased by Ernest Doe & Sons on 1 November, and Colchester and its satellite depot at Marlesford became Doe branches. Following an agreement made on 28 April 1998, giving Doe the New Holland agricultural franchise for most of Norfolk, a further two branches were opened at Fakenham and North Walsham. Stansted was closed in 2000 through rationalisation and because of its awkward position in the centre of the town, but expansion of the professional grass machinery business led to dedicated depots being set up at Esher in Surrey in 1997 and Billingshurst in West Sussex in 2000.

Today, the company is the largest New Holland dealership in the UK and the largest machinery retail business in East Anglia. Ernest Doe & Sons has sixteen branches, employs around 500 people and has a turnover of £65 million. In addition to the New Holland dealership for both agricultural and construction machinery, important franchises are held for Lely, Vicon, TIM, Richard Pearson, Netagco Reekie, Lemken, Shelbourne Reynolds, KRM, Bomford, Knight, Allman, Dowdeswell, Marston, Manitou, O&K and Textron equipment, to name a few.

In January 2001, the company marked the completion of its 102nd year of trading with the introduction of a new identity and logo in the style of 'ernest Doe'. The following month saw the staging of the 41st Doe Show with a record attendance. The showroom at Ulting may still be on the site of the original blacksmith's shop, but the company has come a long, long way. And yet, it retains its traditions and remains a family business; Colin's son, Angus, will be the fifth generation to trade at Doe's Corner.

In January 2001, Ernest Doe & Sons celebrated the completion of its 102nd year of trading with the introduction of a new identity and logo as seen on the bonnet of this New Holland TM135, one of the company's demonstration tractors.

Chapter 6

DOE POWER IN ACTION

A Power Major Triple D at work with a Doe four-furrow reversible plough on Orsett Estate near Grays in Essex. This was a heavy-land farm with mixtures of both Essex and London clay.

952 UPU

Over a third of the Doe tractors sold stayed in Essex. The Triple D was designed to suit the local conditions and was the ideal machine to cope with the heavy Essex clay soil. Most went to large estates and farming companies.

The biggest customer for both Triple D and 130 tractors was Wallasea Farms, based near Southend-on-Sea. This company bought thirteen Triple Ds and several 130s, before moving over to County tractors when Doe ceased production.

Further back down the Thames estuary at Grays, the Orsett Estate grew continuous cereals on 2,600 acres of mainly heavy Essex and London clay soil. A Triple D carried out most of the ploughing until a new 130 was bought in June 1965. This tractor coped admirably with the sticky conditions and clocked up over 2,000 hours without any major mechanical mishaps during its first two seasons.

Lord Rayleigh's Farms covered over 7,000 acres at Terling, Hatfield Peverel and Leighs Priory in the Chelmsford area. The farms were committed Doe users and had several Triple D and 130 tractors, as well as the last 150 model. Field trials for the Triple D were carried out on the estate, and the farms manager, Harold Lepper, and the farm bailiff at Terling Hall, George Cook, were involved in several trials, including the twelve-hour marathon ploughing test held in August 1961.

Essex accounted for over 100 Triple D tractors. No other county came close to matching these figures, although over forty were sold into neighbouring Suffolk, making it the second best sales area followed closely by Lincolnshire.

E.A. Sheardown & Co. Ltd. of Grantham in Lincolnshire replaced its crawlers with two Doe 130s in June 1966. By the following January, the two machines had each done between 1,700 and 1,800 hours' work apiece. During peak cultivating periods, Sheardown's Doe tractors were worked twenty-four hours a day for six days a week with the drivers taking twelve-hour shifts. Fuel consumption worked out at between two and three gallons per hour.

The Doe Triple D and 130s were only really suitable for the large arable farms, and significant sales were made to Hertfordshire, Norfolk, Cambridgeshire, Hampshire and Yorkshire. Sales in other areas of the UK only made single figures.

Nottinghamshire had only one Triple D tractor until ploughing contractor Alwyn Blatherwick bought a second-hand machine in 1964. Blatherwick's machine, built in 1962, more than paid for itself during the first season, ploughing over 800 acres at £2 per acre. Over the next six years, the tractor ploughed between 800 and 1,100 acres a year, often clocking up over 14 hours a day. Alwyn still owns the Triple D, and still goes ploughing with it!

Outside agriculture, only a few Doe tractors were sold for industrial use and these were used mainly for motorway construction work. Of those that were exported, most went onto farms, although one or two were sent to Scandinavia for drainage work. As already mentioned, some were sold to the West Indies for hauling cane trailers on a sugar plantation.

Unfortunately, once the Doe tractor went out of fashion and the concept was overtaken by more modern and more powerful four-wheel drive machines, very many of the Triple Ds and 130s suffered the fate of being broken up or made back into two tractors. This was particularly true of the 130 model, which became worth more as two individual Ford 5000 units that could be sent for export to meet the demand for second-hand tractors from the Middle and Far East during 1980s.

Tractor exporters and dealers Agrimac recall breaking up at least ten Doe 130s at its Arrington premises in Cambridgeshire during the mid-1980s. The Ford 5000 tractor units were all sent to Thailand. Machinery dealers Frank Foot & Son of Winchester in Hampshire had twenty-seven Doe tractors through their hands over the years. Eighteen of these were broken up, the 5000 units also being exported to Thailand. This was the main export market for Ford models, and Frank Foot shipped over five hundred tractors there in 1990. Most of the other exporters and dealers in the UK would have broken at least some Doe tractors, making the 130 now a very scarce commodity.

Luckily, a number still survive and, not surprisingly for such a unique machine, Doe tractors have become highly prized among collectors and attract high prices. Most of those that still exist today have been restored or have gone into the preservation movement. They have a large following and their own club.

The Doe Owners Club is run by Alwyn Blatherwick, the Nottinghamshire ploughing contractor. He became fascinated with the tandem machines after he bought his Triple D in 1964, and has attempted to record all the Doe tractors still in existence. He has over seventy on his register and believes that around 100 still exist. As with most Doe enthusiasts, he likes both to exhibit and work his tractors, and the Triple Ds that were once were surrounded by onlookers at agricultural shows now attract the crowds on the rally field.

A Super Major Triple D working with an early Doe tubular-beam reversible plough on Lord Rayleigh's Farms in 1961. The field was close to the A12 at Hatfield Peverel. Lord Rayleigh's Farms covered over 7,000 acres at Terling, Hatfield Peverel and Leighs Priory.

A Triple D is refuelled on Lord Rayleigh's Farms. Preparing a Doe tractor for the day's work meant filling two fuel tanks and checking the oil and water in two engines.

The Triple D in action on Lord Rayleigh's Farms at Terling Hall. The tractor is pulling a six-furrow Doe plough and has wheel weights all round.

A Power Major Triple D on demonstration with a four-furrow plough in Lincolnshire in 1960. It was unusual to fit steel wheels to the front tractor unit while retaining rubber tyres at the rear - an arrangement that would probably cause some wind-up between the transmissions.

A Doe Triple D that was used in Lincolnshire by Burtts of Dowsby. Based on two Power Major skid units, it is fitted with an early Lambourn cab and is seen working with a four-furrow Ransomes FR TS73 in early 1963.

Nottinghamshire ploughing contractor Alwyn Blatherwick bought this 1962 Triple D when it was two years old. Between 1964 and 1970, it ploughed between 800 and 1,100 acres a year and is seen hard at work with a seven-tine cultivator in 1968.

Triple D and 130 tractors were used on land farmed by the Doe family.
A Super Major Triple D is seen working near Ulting with a Doe toolcarrier and a four-furrow reversible plough.

The few Triple D tractors that were sold for industrial use were bought by contractors involved in road construction.
This 1962 Triple D was used to haul a scraper during the building of the M1 motorway extension into London.

Doe tractors could cope with the worst of all weathers, and this Super Major Triple D was used by a local farmer to clear snow from the roads around Bishop's Stortford in Essex during the severe winter of 1963.

This Triple D, serial No. 313, was one of the first New Performance models. It was supplied new to D. J. Fisher Farms Ltd. on 23 August 1963 and is seen having its mounted plough set up soon after delivery. It was replaced by a new Doe 130 in May 1968.

This New Performance Triple D, believed to be No. 348, was used for trials with the National Institute of Agricultural Engineering at Silsoe. It is seen towing a prototype plastic pipe drainage machine in 1964.

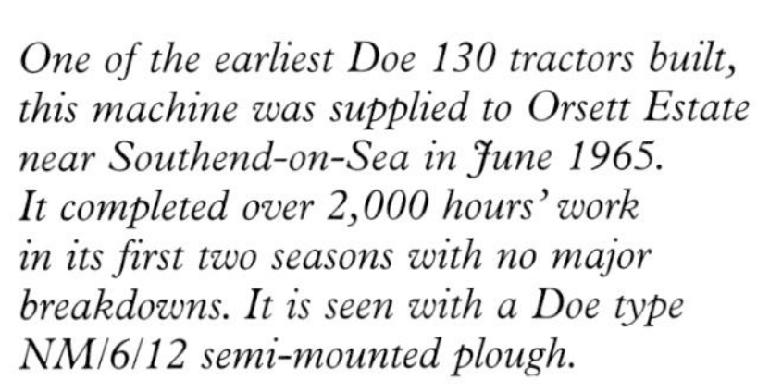

One of the earliest Doe 130 tractors built, this machine was supplied to Orsett Estate near Southend-on-Sea in June 1965. It completed over 2,000 hours' work in its first two seasons with no major breakdowns. It is seen with a Doe type NM/6/12 semi-mounted plough.

One of two Doe 130 tractors purchased in July 1966 by E. A. Sheardown & Co. Ltd. of Grantham, Lincolnshire. By the following January, both machines had each completed between 1,700 and 1,800 hours' work on cultivating and pea vining.

This Doe 130 was one of a batch of eight exported to Nigeria in 1965 for work on a sugar plantation. It was used to haul cane trailers and transport the crop from the field to the refinery at harvest time. Note the extended frame on the rear unit to accommodate the different support mountings for the front tractor to allow greater ground clearance.

This industrial Doe 130 was supplied new to Fitzpatrick in 1965 and was used for road and motorway construction work.

A 1966 Doe 130 at the end of its working days. By the late-1970s, the tandem tractor concept had become outdated and was overtaken by more modern four-wheel drive machines. Many Doe 130 tractors were broken up during the mid-1980s and their Ford 5000 units sent for export.

This Doe 130 appeared in a farm auction near Kettering in Northamptonshire during the early 1980s. Apparently, it did not make its £1,000 reserve and was rumoured to have been destroyed by fire soon after the sale. The Lambourn safety cab seems to have been a popular later fitment on many Doe tractors.

Alwyn Blatherwick, the enthusiast who runs the Doe Owners Club. He is an expert on Triple D tractors and has owned one since 1964.

Typical of the many Doe tractors in preservation, this 130 is owned by Paul Yates of Gloucestershire. It was originally one of the two tractors operated by E. A. Sheardown of Grantham and is pictured on page 100. It is shown at the Tractor Millennium, held at Newark in 2000, where it formed part of one of the largest gatherings of Doe tractors ever seen on a rally field.

Appendix 1 - Colchester Tillage

Doe's Colchester branch was opened in 1943 after Ernest Charles Doe acquired the premises at 79 Hythe Hill from the coal merchants, Thomas Moy. It operated as a branch of Ernest Doe & Sons until the Fordson franchise for the Colchester area became available after the local dealer, Willett, went into liquidation in 1946. Doe was keen to take over the franchise, but the Ford Motor Company was reluctant to award it to a company that was selling competing makes of tractor.

Ford went as far as suggesting that if Doe wanted the Fordson franchise for the Colchester area, it should give up selling other makes of tractors at all its branches, which was of course completely unacceptable to Ernest Charles Doe. A compromise was eventually reached in that Colchester Tractors Ltd was formed as a separate company with effect from 19 May 1947 to take over the Fordson franchise and operate out of the Hythe Hill premises. The directors of the new company were Ernest Charles Doe and Herbert Walter Doe with Stephen Woodridge joining them as managing director.

In 1952, Alan Doe became managing director of Colchester Tractors, a position that he held until 1957 when he was succeeded by Owen Dunt. Dunt's time at Hythe Hill was marked by

A Colchester Tractors brochure showing the range of equipment it handled in 1959. The company imported Flemstofte, Lely, Lemken and Hammerbo machinery.

In 1960, Colchester Tillage was appointed the sole UK concessionaires for the Tico hydraulic lorry cranes imported from Sweden. This 1 ton Mini-Crane was designed for unloading 10 to15 cwt pick-ups and is seen fitted to a Ford 400E truck.

Colchester Tillage introduced its first Cotil tractor-mounted crane, designed to fit Fordson Power Major or Super Major tractors, in 1961. For more robust applications, it brought out this crawler crane, based on a County Mk lV industrial crawler. It was ideal for pipe-laying applications.

the establishment of the company as sole concessionaires for a variety of imported equipment. This sideline began with agencies for the Flemstofte harrow, the Lely precision broadcaster, the Hammerbo manure spreader and Lemken reversible ploughs. The German plough range, launched in the UK at the 1959 Royal Show, became the company's most successful concession.

In 1960, chaiman Alan Doe set up Colchester Tillage as a wholly owned subsidiary of Colchester Tractors to handle the importation of farm equipment from Europe, Scandinavia and the USA. Several other concessions, including Bjurenwall, Gandy and Kemper, were added.

The following year, Colchester Tillage was appointed sole UK concessionaires for the Tico hydraulic lorry crane from Sweden. Clamped to the chassis, it could be used to load or unload the vehicle and was powered by a hydraulic pump. The crane featured a telescopic jib incorporating a hydraulically-driven winch with a steel cable to haul loads from inaccessible sites.

The company also adapted theTico K52E crane to fit Fordson Power Major and Super Major tractors. The crane's pedestal allowed it to slew

The Cotil CT74 crane, based on a Ford 5000, was introduced in 1966. It was priced at £1,840 to include the tractor unit equipped with 14 x 30 tyres.

A Cotil CT74 showing the crane in the folded transport position.

through 360 degrees and was centrally-mounted over the bonnet and supported by a sub-frame attached to the tractor's side-channels. Launched at the 1961 Smithfield Show as the Cotil tractor crane, the attachment cost £450.

The Cotil CT74 1 ton crane based on the Ford 5000 tractor was introduced in 1966. It had a 13 ft telescopic boom and cost £1,840 to include the tractor unit equipped with 14 x 30 tyres. The crane proved to be popular on building sites and for light industrial use. A rear-mounted version, the CT74RM, was introduced during the 1970s.

Colchester Tillage marketed the CT300 and CT450 cranes through the 1980s, based on both Ford 6600 and 6610 tractors. The CT300 was a 1 ton crane while the CT450 would lift 2 tons. The CT450 could also be equipped with hydraulic attachments such as a grab, rotator or clamshell bucket.

For more robust applications, the company built a number of cranes on County tractors and enjoyed a long association with this well-known four-wheel drive manufacturer. The first County application, a crane fitted to an industrial crawler

The Cotil CT74 had a 1 ton lift capacity and a 13 ft telescopic boom. It was ideal for light industrial use or work on a building site.

A Cotil CT5004 crane based on a County 1124 tractor. The County Cotil conversions found many applications within the forestry and construction industries, both in the UK and overseas. They were used as far afield as the Middle East, West Indies and the Arctic Circle.

during the 1960s, was followed by the four-wheel drive Cotil CT5004 based on an 1124 tractor. This was replaced by the CT4504, a 2.5 ton capacity machine using County 6600-Four, 1174 or 1184TW skid units.

The County conversions found many applications with the forestry and construction industries both in the UK and overseas. They were used for laying oil pipes in the Middle East, loading sugar cane in the West Indies and laying railway lines near the Arctic Circle in Finland. The last Cotil County appeared in 1988 and was

The Cotil CT5004 demonstrates its reach. It had a 2.5 ton capacity. Note that the tractor is fitted with a Wickham-Poole hitch for handling semi-trailers.

This impressive beast is a County 1174 fitted with a later Cotil CT4504 crane and a rear-mounted Sanderson fork-lift attachment. It was used by a plant-hire firm in Wales.

based on an 1164TW tractor fitted with the latest CT772RM rear-mounted crane.

Following the sale of a significant part of Colchester Tillage's business to Lemken UK in 1993, it was felt that the cost of the Colchester group of companies, still controlled by the Doe family, remaining independent was too high. It was a logical step to bring the group back under the wing of Ernest Doe

A rear-mounted version of the Cotil crane, the CT450RM, is seen fitted to a County 1184TW tractor that was used by Wimpey Mining.

This must rate as the ultimate machine, equipped for anything! It is a Cotil CT4504 2.5 tonne crane, based on a County 1184TW fitted with a McConnel backhoe and a front-mounted hydraulic winch.

& Sons, which effectively purchased Colchester Tractors and Colchester Tillage on 1 November 1993 after a deal was agreed by the two boards.

Colchester Tractors was reorganised as a branch of Ernest Doe & Sons, back in Colchester after nearly fifty years, while Colchester Tillage operated as a trading division of the company, importing Italian Cormach lorry cranes along with other parts from past Tillage franchises. The Cotil business continues and a batch of three cranes for export was completed in 1995. Built for the gas board in Pakistan, they were based on four-wheel drive Ford 8240 skid units fitted with 2.5 tonne Cormach cranes. The current range features cranes from 1 to 3.5 tonne lifts.

Built for export in 1995, this machine represents one of the latest Cotil crane conversions. It is based on a Ford 8240 skid unit fitted with a 2.5 tonne Cormach crane.

Appendix 2 - Tandem Tractors

As remarkable as they were, the Triple D and its sister machines were not unique, as other tandem tractor designs were produced both in the UK and other parts of the world. Although Doe received accolades for its dual tractors, the company was not the first to pioneer the concept of joining two skid units together to provide a more powerful machine with greater traction.

The tandem tractor concept seems to have originated in the USA and gained recognition during the early 1950s after a number of Midwest farmers linked two of the big single-cylinder John Deere R tractors together to make 100 hp machines. Pioneering work carried out in 1955 by the Iowa Agricultural Experimental Station saw a Ford 8N coupled to a Ford 860 tractor and the unit steered by the independent brakes. This venture was developed further by Professor Wesley Buchels of the agricultural engineering department of Michigan State University.

Buchels' work at Michigan resulted in a more sophisticated tandem tractor, based on two gasoline Ford 960-S skid units and built with input from Ford's Tractor & Implement Division. The machine was exhibited at the American Society of Engineers' fiftieth anniversary conference in September 1957 and received a great deal of media attention. The tractors were joined

This tandem tractor was developed Professor Wesley Buchels of the agricultural engineering department of Michigan State University. Based on two gasoline Ford 960-S skid units, it was built in conjunction with the American Ford Tractor & Implement Division and is seen on demonstration in 1959.

Built in Australia, the Hewco Twin Powered Scraper looks slightly reminiscent of the Triple D, but there all similarity ends. It was powered by two Fordson Power Major tractor skid units working in tandem, one pulling while the other pushed.

The first MIAK conversion was based on two Nuffield 4/60 tractors supplied through BMC Sweden AB in Stockholm. It is believed that only the one machine was built.

by a welded sub-frame that used a tongue and king-pin arrangement to provide the articulation. It differed from the Iowa design in that it was steered by two single-acting hydraulic rams operated by the 960's original power steering equipment. Dual controls for the throttle, starting and ignition were located on the rear tractor unit with a master cylinder system, similar to that used on the Doe, to actuate the clutch.

The Michigan State University tractor evidently performed well in tests and the design was found to provide excellent weight distribution across the four driving wheels. Contemporary reports suggest that interest was shown by both the Ford and Caterpillar organisations, but it seems that no serious attempts were made to take the design any further.

Australia had its share of tandem tractor developments, and one of the first applications of the concept was little more than an improvisation carried out in the farm workshop. The machine, built in 1953 and based on two British Field Marshall tractors, was the work of the Wilkins brothers who farmed in Western Australia. The rear tractor had its front axle removed and was fitted with a pivot pin that fixed to the drawbar of the front tractor. The front unit retained its front axle.

The driver controlled the dual tractors from the front seat, with the clutch, governor and throttle on the rear

Later MIAK conversions were based on BM Volvo 350 tractors with 35 hp three-cylinder diesel engines. Only a limited number of the machines were made due to lack of interest in the country.

A MIAK BM Volvo conversion on demonstration in Sweden with two ploughs also coupled in tandem.

unit operated by a system of control rods and bell-cranks. The brothers claimed that the tandem machine would outperform a crawler, but admitted that the noise and vibrations set up by the two single-cylinder engines running in close proximity could be a problem. In the words of one of the Wilkins brothers, 'The drawbar of the front tractor had to be spring loaded, as when both engines are not firing at exactly the same time, the pull is jerky.' Anyone who has had any experience of Field Marshalls will know exactly what he meant!

There were several other tandem tractor conversions in Australia. Plant & Plant of Queensland linked together two Case 832C tractors equipped with torque converter transmissions, while Harts Machinery of Parkes, New South Wales, built a similar machine to the Doe 130 out of two Ford 5000 units. The Pederick 'Dual Power', made from two Chamberlain Champion 306 skid units by Pederick Engineering of Western Australia, appeared in 1969. The closest relative to the Triple D was the Hewco Twin Powered Scraper, which was based on two Fordson Power Major tractor units working in tandem.

In Europe, A. H. Steenburger of the Netherlands joined two International 436 tractors, while the French Le Bisom-Trac was a 90-hp tandem tractor, similar in appearance to the Doe, that was produced in 1956 from two 45-hp Someca 40H skid units. Experiments were even carried out in Russia with two 21-hp Vladimir T28 tractors coupled in tandem.

The closest rival to the Triple D was the MIAK conversion produced in Sweden. The first of these was based on two Nuffield 4/60 tractors supplied through BMC Sweden AB in Stockholm. Later MIAK conversions used BM Volvo 350 tractors with 35 hp three-cylinder diesel engines. A limited number of the machines were made before production was eventually stopped due to lack of interest in Sweden.

Doe had little competition in the UK. A Berkshire farmer, Pat Saunders, experimented with linking a Fordson Major to a Nuffield Universal to gain more power and extra traction. His developments led to a kit consisting of a sub-frame that supported the rear unit. All the controls were grouped at the front of the sub-frame that also had a ring hitch to attach it to the pick-up hitch on the leading tractor. From 1968, Saunders' system was marketed by Paramount Engineering of Coventry as the Dual Tractor Kit. Priced at £160, it was advertised as suitable for coupling any two popular makes of tractor. It was demonstrated with either an International 634 or a Ford 3000 linked to a Fordson Major, but drew little interest.

The wet autumn of 1968 prompted Suffolk farmer Adrian Sampson to devise a hitch to couple his David Brown 990 and 1200 tractors together to complete the ploughing on his 365 acre holding. The hitch allowed the front tractor linkage to take the weight of the rear tractor with its front-axle extensions removed. A high-

Paramount Engineering's Dual Tractor Kit was designed to link any two popular makes of tractor together, in this case a Ford 3000 and a Fordson Super Major. The kit consisted of a sub-frame that supported the rear unit, with all the controls grouped within reach of the driver on the front tractor.

The sub-frame supporting the rear tractor on Paramount's kit had a ring hitch to attach it to the pick-up hitch on the leading tractor. If the hitch became worn, it was not unknown for the two tractors to part company and the rear unit run away under its own power!

Steiger in the USA went one better than the tandem and brought out this tricycle arrangement in the mid-1970s. Nicknamed 'Big Jack', it was based on three Cougar tractors and had a massive combined output of over 750 hp.

tensile steel rod and two circular steel plates provided the connection and articulation between the two tractors.

The advantage of the system was its simplicity. The leading tractor's lift could also be used to raise the front of the rear tractor off the ground for the removal of the front wheels and axle extensions. Sampson planned to market the conversion as the Samco system hitch, priced at £30 with kits available for most standard makes of tractor. Launched in 1970, it met with little commercial success, however.

The ultimate tandem tractor appeared in 1976. It was a 650 hp machine based on two Steiger Panther lll tractors. But this American company had already gone one stage better, because the previous year Steiger's experimental department had introduced 'Big Jack'. This was built out of the front halves of three Cougar ll frames in a tricycle arrangement. The three Caterpillar 3306 engines blasted out a massive 750 hp. As impressive as it was, it could never match the success of the Triple D.

Doe's machines seem to be the only tandem tractors that went into any volume production. It could be argued that their success was partly down to the choice of power unit, as the Ford tractor was accepted the world over. But the real secret of the Triple D was that it was a design that worked; it was particularly suited to the farming conditions of its local area and had the reputation and resources of a respected family company behind it.

When the going got tough … a Yorkshire farmer coupled two David Brown 990 tractors together in tandem to cope with winter ploughing in January 1967. The combined power of the two 55hp engines gave him over 100hp to play with.

Appendix 3 - Doe Triple D & 130 Tractors - Numbers Sold

	Doe Triple D							Doe 130	
Home	**1958**	**1959**	**1960**	**1961**	**1962**	**1963**	**1964**	**1965**	**Total**
Bedfordshire		1		1		2			4
Berkshire						2		2	4
Cambridgeshire			1	2	1	2	1	4	11
Derbyshire							1		1
Dorset				1					1
Essex	1	5	11	23	17	37	10	19	123
Hampshire						1	2	8	11
Hertfordshire		2	3	3	5	4	1	4	22
Huntingdonshire						1		1	2
Kent			2	1			1		4
Lancashire							1		1
Leicestershire				1	1	1		1	4
Lincolnshire			2	3	12	6	4	4	31
Middlesex						1	1		2
Norfolk					4	5	1	2	12
Northamptonshire						1		1	2
Nottinghamshire		1			1		1		3
Oxfordshire						3	3		6
Somerset						1			1
Suffolk	2	1	4	12	10	5	8	6	48
Sussex							1	1	2
Warwickshire				1		2	2	1	6
Yorkshire				1	2	1	2	4	10
Scotland						1		1	2
Total home sales	3	10	23	49	53	76	40	59	313
Export									
Australia								1	1
Austria							1		1
Denmark							1	2	3
Germany				1	1	3			5
Ireland			1						1
Israel						1			
Nigeria					1			8	9
South Africa					1	1			2
Southern Rhodesia							1		1
Spain					1				1
Sweden		1	2	3				3	9
Uruguay				1					1
USA				1	3				4
USSR		2					1		3
Total export sales	0	3	3	6	7	5	4	14	42
Overall Total	**3**	**13**	**26**	**55**	**60**	**81**	**44**	**73**	**355**

NB: These numbers may not be complete as some records were destroyed by fire

Bibliography

Journals & Periodicals

Doe News
Farm Implement & Machinery Review
Farm Mechanisation
Farmer & Stockbreeder
Power Farming

NIAE Test Reports: No.396/BS; No.408; No.413/O.E.C.D.

Books

Doe, Alan, A *Century of Service* (Ernest Doe & Sons Limited 1998)
Gibbard, Stuart, *Ford Tractor Conversions* (Farming Press 1995)
Leffingwell, Randy, *Ford Farm Tractors of the 1950s* (MBI Publishing Company 2001)

Useful Addresses

Ernest Doe & Sons Ltd.
Ulting
Maldon
Essex
CM9 6QH
Telephone: 01245 380311 Website: www.ernestdoe.com

Doe Owners Club
Alwyn Blatherwick
Moor Lane
South Clifton
Newark
Nottinghamshire
NG23 7AN
Telephone: 01522 778486

Index

Numbers in bold type refer to illustrations

F

G

H

I

J

K

L

M

N

O

P

R

S

T

V

W

Y

About the Author

A successful author and journalist specialising in tractors and machinery, Stuart Gibbard comes from a farming background near Spalding in Lincolnshire. He has developed his interest in collecting early tractor literature into a mail-order business which is run by his wife Sue, and he is also one of the organisers of the annual Spalding model-tractor and literature show.

Stuart's first book, published to great acclaim in 1994, was *Tractors at Work*, a remarkable collection of rare and archive photographs spanning ninety years of tractor development from Dan Albone's 1904 Ivel. Stuart has since produced another nine tractor books, including prize-winning books on Ford tractors, and four videos.

Devoting much of his time to historical research, Stuart has talked to many of the men who played their part in creating the machines portrayed in the books. This first-hand knowledge has enabled him to give a fascinating insight into the world of agricultural engineering and tractor development, and his publications include unrivalled histories of County, Roadless and other Ford tractor conversions.

Previous publications

Books

Tractors at Work: a pictorial review 1904-94 (1994).

Ford Tractor Conversions: the story of County, Doe, Chaseside, Northrop, Muir-Hill, Matbro and Bray (1995).

Tractors at Work: a pictorial review Volume Two (1995).

Roadless: the story of Roadless Traction from tracks to tractors (1996).

Change on the Land: a hundred years of mechanised farming (1997).

County: a pictorial review (1997).

The Ford Tractor Story Part One: Dearborn to Dagenham 1917 to 1964 (1998).

The Ford Tractor Story Part Two: Basildon to New Holland 1964 to 1999 (1999).

The Ferguson Tractor Story (2000).

Videos

Roadless Tractors (1996).

County Tractors (1997).

Ferguson Tractors (1998).

Ferguson on the Farm Part One (2001).